Checklisten der Fauna Österreichs, No. 5

Erhard CHRISTIAN:
Protura (Insecta)

Christian KOMPOSCH:
Opiliones (Arachnida)

VOLKER MAHNERT:
Pseudoscorpiones (Arachnida)

PETER VOGTENHUBER:
Tipulidae (Insecta: Diptera)

Herausgegeben von Reinhart Schuster

Serienherausgeber
Hans Winkler & Tod Stuessy

Verlag der
Österreichischen Akademie
der Wissenschaften

Wien 2011

Titelbild: *Megabunus armatus* (KULCZYNSKI, 1887). – Ein als Endemit der Südlichen Kalkalpen zu wertender Weberknecht, der mit seiner Körperfärbung sehr gut an die von ihm bewohnten Kalkfelswände angepasst ist; Vellacher Kotschna, Steiner Alpen; (Foto: Ch. KOMPOSCH)

Layout & technische Bearbeitung: Karin WINDSTEIG

Checklists of the Austrian Fauna, No. 5. Erhard CHRISTIAN: Protura (Insecta), Christian KOMPOSCH: Opiliones (Arachnida), Volker MAHNERT: Pseudoscorpiones (Arachnida), Peter VOGTENHUBER: Tipulidae (Insecta: Diptera).

ISBN 978-3-7001-7052-5, Biosystematics and Ecology Series No. 28, Austrian Academy of Sciences Press; volume editor: Reinhart SCHUSTER, Institute of Zoology, Karl-Franzens-University, Universitätsplatz 2, A-8010 Graz, Austria; series editors: Hans WINKLER, Commission for Interdisciplinary Ecological Studies, A-1010 Vienna, Dr. Iganz Seipel-Platz 2, Austria & Tod STUESSY, Institute of Botany, University of Vienna, A-1030 Vienna, Rennweg 14, Austria.
A publication of the Commission for Interdisciplinary Ecological Studies (KIÖS)

Checklisten der Fauna Österreichs, No. 5. Erhard CHRISTIAN: Protura (Insecta), Christian KOMPOSCH: Opiliones (Arachnida), Volker MAHNERT: Pseudoscorpiones (Arachnida), Peter VOGTENHUBER: Tipulidae (Insecta: Diptera).

ISBN 978-3-7001-7052-5, Biosystematics and Ecology Series No. 28, Verlag der Österreichischen Akademie der Wissenschaften; Bandherausgeber: Reinhart SCHUSTER, Institut für Zoologie, Karl-Franzens-Universität, Universitätsplatz 2, A-8010 Graz, Österreich; Serienherausgeber: Hans WINKLER, Kommission für Interdisziplinäre Ökologische Studien, A-1010 Wien, Dr. Ignaz Seipel-Platz 2, Österreich & Tod STUESSY, Institut für Botanik, Universität Wien, A-1030 Wien, Rennweg 14, Österreich.
Eine Publikation der Kommission für Interdisziplinäre Ökologische Studien (KIÖS)

Anschrift der Verfasser:
Univ.-Prof. Dr. Erhard CHRISTIAN, Institut für Zoologie, Department für Integrative Biologie, Universität für Bodenkultur, Gregor-Mendel-Straße 33, A-1180 Wien, Österreich – Email: erhard.christian@boku.ac.at
Dr. Volker MAHNERT, c/o Muséum d'histoire naturelle, 1, route de Malagnou, CH-1208 Genf, Schweiz – Email: volker.mahnert@wanadoo.fr
Mag. Dr. Christian KOMPOSCH, ÖKOTEAM – Institut für Tierökologie und Naturraumplan-ung, Bergmanngasse 22, A-8010 Graz, Österreich – Email: c.komposch@oekoteam.at
Dipl.-Ing. Peter VOGTENHUBER, Biologiezentrum, J.-W. Klein-Str. 73, A-4040 Linz, Österreich – Email: p.vogtenhuber@landesmuseum.at

©2011 Austrian Academy of Sciences
Printed in Austria by Riegelnik

Inhalt

Peter Vogtenhuber

Tipulidae (Insecta: Diptera)

Protura (Insecta)

Erhard CHRISTIAN

Summary: Currently 58 species of Protura are reliably recorded from Austria. Records of another eight species are doubtful.

Zusammenfassung: Derzeit sind aus Österreich 58 Proturenarten gesichert. Nachweise von weiteren acht Arten sind bestätigungsbedürftig.

Key Words: Protura, Austria, checklist, biodiversity

I Einleitung

Protura (Beintastler) sind eine Gruppe primär flügelloser Insekten mit entognathen Mundwerkzeugen, deren Monophylie nie bezweifelt wurde. Ein markantes Eigenmerkmal – der funktionale Ersatz der Antennen durch das erste Laufbeinpaar – macht die blinden Bodenschlängler unverkennbar. Die lange Zeit vorherrschende Ellipura-Hypothese postuliert ein Schwestergruppenverhältnis zwischen Protura und Collembola. Sie wird durch neuere Untersuchungsergebnisse erschüttert, die eine gemeinsame Wurzel von Protura und Diplura andeuten (Giribet et al. 2004). Das Verzweigungsmuster an der Basis der Hexapoda ist nach wie vor umstritten.

Als letzte der artenreichen Insektenordnungen wurden die Proturen erst 1907 entdeckt. Der maximal zwei Millimeter lange und in Anpassung an das Leben im Boden wurmförmig gestreckte Körper trägt dazu bei, dass Proturen oft übersehen werden. In der gestaltlichen Vielfalt reichen sie an ähnlich kleine Bodenarthropoden wie Milben oder Collembolen nicht heran. Selbst die hochrangigen Untergruppen differieren nur in unauffälligen morphologischen Details, und zur Artbestimmung müssen chaetotaktische Merkmale geprüft werden, die auch ein geübter Mikroskopiker erst nach längerer Lehrzeit beurteilen kann. Die regional sehr ungleichmäßige Kenntnis der Proturenfauna hängt mit diesen Schwierigkeiten zusammen: Determinatoren sind rar und auf Jahre hinaus ausgebucht.

II Allgemeiner Teil

1. Erforschungsgeschichte und aktueller Forschungsstand in Österreich

Österreich war bis in die 1950er-Jahre ein weißer Fleck auf der Proturenlandkarte. Aus der Zeit davor existieren nur wenige punktuelle Meldungen, von denen eine einzige, die älteste, anhand eines Belegexemplars überprüft werden konnte. Jan STACH, der große polnische Apterygotenforscher der ersten Jahrhunderthälfte, hat am 13.05. 1915 in Wien-Neuwaldegg jenes Exemplar gesammelt, das er nach dem Krieg unter dem Namen des klassischen Proturen *Acerentomon doderoi* SILVESTRI, 1907 veröffentlichte (STACH 1926). Es handelt sich um *Acerentomon dispar*, eine von STACH selbst, allerdings viel später beschriebene Art (SZEPTYCKI in litt.). Umfangreichere faunistische Angaben lieferte erst Herbert FRANZ mit den Bestimmungen seiner Mitarbeiterin Elisabeth SERTL-BUTSCHEK (FRANZ & SERTL-BUTSCHEK 1954). Zwischen 1965 und 1977 publizierte der in Pressburg tätige Taxonom Josef NOSEK wesentliche Beiträge zur österreichischen Proturenfauna. Er kooperierte mit den österreichischen Zoologen Konrad THALER (Innsbruck; s. THALER 1994), Herbert FRANZ (Wien; s. FRANZ et al. 1969) sowie mit Reinhart SCHUSTER und Heinz NEUHERZ (beide Graz; s. NOSEK 1977b, NEUHERZ & NOSEK 1975). Nach zwei Jahrzehnten Stillstand brachte Andrzej SZEPTYCKI (Krakau) neuen Schwung in die österreichische Proturenfaunistik.

Seine Determinationen und Neubeschreibungen führten u.a. zur Aufdeckung der erstaunlichen Proturendiversität im Wiener Stadtgebiet (CHRISTIAN & SZEPTYCKI 2004).

III Spezieller Teil

Bei derzeit 748 anerkannten Arten (SZEPTYCKI 2007a) liegt der österreichische Anteil an der globalen Proturenfauna knapp unter acht Prozent. Von den für Europa verbuchten Arten (SZEPTYCKI 2007b) wurde jede dritte auch oder nur in Österreich gefunden. Die hohe apparente Artdichte ist nicht mit überdurchschnittlichem Erfassungsaufwand zu erklären. Fundpunkte häufen sich lediglich in den Aktionsräumen einiger Bodenzoologen, während Gebiete von der Größe des Bundeslandes Vorarlberg nahezu blank sind.

In Österreich wurden 58 Proturenarten vertrauenswürdig nachgewiesen, acht weitere bestätigungsbedürftige Arten werden in der Liste ignoriert (aber im Abschnitt "Problematica" angeführt). Die österreichischen Arten verteilen sich auf zehn Gattungen, vier Familien und zwei Unterordnungen.

Die Nomenklatur richtet sich nach dem Katalog der Proturen der Welt (SZEPTYCKI 2007a) und nach der "Fauna Europaea" (SZEPTYCKI 2007b). Bundesländer werden mit den alphabetisch angeordneten Kürzeln bezeichnet:

Kürzel der Bundesländer:

B	=	Burgenland
K	=	Kärnten
N	=	Niederösterreich
O	=	Oberösterreich
S	=	Salzburg
St	=	Steiermark
T	=	Tirol
V	=	Vorarlberg
W	=	Wien

Österreichische Typlokalitäten sind angegeben.

1. Liste der in Österreich nachgewiesenen Arten

Unterordnung ACERENTOMATA

Familie Hesperentomidae

Gattung *Ionescuellum* TUXEN, 1960

Ionescuellum condei (NOSEK, 1965)
    N (loc. typ.: Mödling, Frauenstein: NOSEK 1965a; sub *Hesperentomon*), W
Ionescuellum haybachae (NOSEK, 1967)
    N (loc. typ.: Steinwandklamm bei Furth: NOSEK 1967; sub *Hesperentomon*)
Ionescuellum schusteri (NOSEK, 1977)
    St (loc. typ.: Arzberggraben bei Übelbach: NOSEK 1977b; sub *Hesperentomon*)
Ionescuellum silvaticum (RUSEK, 1965)
    W
Ionescuellum ulmiacum RUSEK & STUMPP, 1989
    W

Familie Protentomidae

Gattung *Protentomon* EWING, 1921

Protentomon tuxeni NOSEK, 1966
    N (loc. typ.: Mödling, Frauenstein: NOSEK 1966)

Gattung *Proturentomon* SILVESTRI, 1909

Proturentomon condei NOSEK, 1967
    T (loc. typ.: Martinswand bei Innsbruck: NOSEK 1967), W
Proturentomon discretum CONDÉ, 1961
    W
Proturentomon minimum (BERLESE, 1908)
    B, N, W
Proturentomon noseki RUSEK, 1975
    W
Proturentomon picardi CONDÉ, 1960
    W
Proturentomon pilosum RUSEK, 1975
    W

Familie Acerentomidae

Gattung *Acerella* BERLESE, 1909

Acerella muscorum (IONESCO, 1930)
K, N, St, W

Gattung *Acerentomon* SILVESTRI, 1907

Acerentomon affine BAGNALL, 1912
N
Acerentomon dispar STACH, 1954
N, O, St, W
Acerentomon fageticola RUSEK, 1966
W
Acerentomon franzi NOSEK, 1965
N, O (loc. typ.: Rosenhof bei Sandl: NOSEK 1965b), W
Acerentomon gallicum IONESCO, 1933
N, St, W
Acerentomon imadatei NOSEK, 1967
K
Acerentomon maius BERLESE, 1908
O, St
Acerentomon meridionale NOSEK, 1960
B
Acerentomon microrhinus BERLESE, 1909
St (Determination unsicher, siehe NOSEK 1973), W
Acerentomon nemorale WOMERSLEY, 1927
St
Acerentomon pseudomicrorhinus NOSEK, 1977
St (loc. typ.: Arzberggraben bei Übelbach: NOSEK 1977a)
Acerentomon quercinum IONESCU, 1932
N, O, W
Acerentomon tuxeni NOSEK, 1961
N, O, S, St, W

Gattung *Acerentulus* BERLESE, 1908

Acerentulus exiguus CONDÉ, 1944
St, W
Acerentulus gisini CONDÉ, 1952
O, St, W
Acerentulus insignis CONDÉ, 1945
W

Acerentulus ruseki Nosek, 1967
    N (loc. typ.: Mödling, Frauenstein: Nosek 1967)
Acerentulus silvanus Szeptycki, 1991
    W
Acerentulus traegardhi Ionesco, 1937
    K, N, W
Acerentulus tuxeni Rusek, 1966
    W

Gattung *Berberentulus* Tuxen, 1963

Berberentulus berberus (Condé, 1948)
    W

Gattung *Gracilentulus* Tuxen, 1963

Gracilentulus fjellbergi Szeptycki, 1993
    W
Gracilentulus gracilis (Berlese, 1908)
    N, St, W

Gattung *Vindobonella* Szeptycki & Christian, 2001

Vindobonella leopoldina Szeptycki & Christian, 2001
    W (loc. typ.: Leopoldsberg: Szeptycki & Christian 2001)

Unterordnung EOSENTOMATA
Familie Eosentomidae

Gattung *Eosentomon* Berlese, 1908

Eosentomon armatum Stach, 1926
    O, S, St, W
Eosentomon bloszyki Szeptycki, 1985
    W
Eosentomon cetium Szeptycki & Christian, 2000
    W (loc. typ.: Leopoldsberg: Szeptycki & Christian 2000)
Eosentomon delicatum Gisin, 1945
    N, O, St, W
Eosentomon gisini Nosek, 1967
    N, St
Eosentomon longisquamum Szeptycki, 1986
    W

Eosentomon luxembourgense SZEPTYCKI, 2001
 W
Eosentomon mariae SZEPTYCKI, 1986
 W
Eosentomon mirabile SZEPTYCKI, 1984
 W
Eosentomon parvum SZEPTYCKI, 1986
 W
Eosentomon pastorale SZEPTYCKI, 2001
 W
Eosentomon pinetorum SZEPTYCKI, 1984
 W
Eosentomon posnaniense SZEPTYCKI, 1986
 W
Eosentomon pratense RUSEK, 1973
 W
Eosentomon rusekianum STUMPP & SZEPTYCKI, 1989
 W
Eosentomon stachi RUSEK, 1966
 T, W
Eosentomon stumppi RUSEK, 1988
 W
Eosentomon transitorium BERLESE, 1908
 B, K, N, O, St, W
Eosentomon vindobonense SZEPTYCKI & CHRISTIAN, 2000
 W (loc. typ.: Leopoldsberg: SZEPTYCKI & CHRISTIAN 2000)
Eosentomon vulgare SZEPTYCKI, 1984
 W
Eosentomon weinerae SZEPTYCKI, 2001
 W

2. Problematica

Von acht weiteren Proturenarten liegen ältere unbestätigte Meldungen aus Österreich vor. Es ist ungewiss, ob sich die publizierten Namen auf die Taxa im heutigen Sinn beziehen. Die folgenden Arten sollten daher bis zur Verifizierung ihres Auftretens in Österreich nicht zur heimischen Fauna gezählt werden.

Ionescuellum montanum (GISIN, 1945)
 St (sub *Proturentomon*); = *I. schusteri*?
Acerella remyi (CONDÉ, 1944)
 O, St (sub *Acerentulus remyi* var. *filisensillatus*)

E. Christian

Acerella tiarnea (BERLESE, 1908)
N (sub *Acerentulus tiarneus*)
Acerentomon doderoi SILVESTRI, 1907
N, O, St
Acerentomon giganteum CONDÉ, 1944
W
Acerentulus confinis (BERLESE, 1908)
B, K, N, O, St,W
Eosentomon germanicum PRELL, 1912
St (sub *Eosentomon spinosum* STRENZKE, 1942)
Eosentomon mixtum CONDÉ, 1945
B, K, N, O, St, W

IV Literatur

CHRISTIAN, E. & SZEPTYCKI, A. 2004: Distribution of Protura along an urban gradient in Vienna. — Pedobiologia **48**: 445–452.

FRANZ, H., HAYBACH, G. & NOSEK, J. 1969: Beitrag zur Kenntnis der Proturenfauna der Nordostalpen und ihres Vorlandes. — Verh. zool.-bot. Ges. Wien **108/109**: 5–18.

FRANZ, H. & SERTL-BUTSCHEK, E. 1954: Protura. — In FRANZ, H.: Die Nordost-Alpen im Spiegel ihrer Landtierwelt. Eine Gebietsmonographie. **Bd.1**, Universitätsverlag Wagner, Innsbruck: 642–643.

GIRIBET, G., EDGECOMBE, G.D., CARPENTER, J.M., D´HAESE, C.A. & WHEELER, W.C. 2004: Is Ellipura monophyletic? A combined analysis of basal hexapod relationships with emphasis on the origin of insects. — Organisms, Diversity & Evolution **4**: 319–340.

NEUHERZ, H. & NOSEK, J. 1975: Zur Kenntnis der Proturen aus dem Gebiet des Faaker Sees in Kärnten. — Carinthia II **165/85**: 297–301.

NOSEK, J. 1965a: A new species of Protura from Austria, *Hesperentomon condei* n.sp. — Rev. Écol. Biol. Sol **2**: 281–283.

NOSEK, J. 1965b: A new species of Protura *Acerentomon franzi* n.sp. — Annotationes zool. et bot. (Bratislava) **14** : 1–4.

NOSEK J., 1966: A new species of Protura from Central Europe *Protentomon tuxeni* sp.n. — Acta soc. zool. Bohemoslov. **30**: 49–53.

NOSEK, J. 1967: The new species of Protura from Central Europe. — Z. Arbeitsgem. österr. Entomol. **19**: 76–88.

NOSEK, J. 1973: The European Protura. Their taxonomy, ecology and distribution with keys for determination. — Muséum d´Histoire Naturelle, Genève, 345 pp.

NOSEK, J. 1977a: A new Proturan species from Styria *Acerentomon pseudomicrorhinus* sp.n. — Rev. suisse Zool. **84**: 345–347.

NOSEK, J. 1977b: *Hesperentomon schusteri* sp.n. a new Proturan species from Austria. — Rev. Écol. Biol. Sol **14**: 593–595.

STACH, J. 1926: *Eosentomon armatum* n. sp., pierwsza Protura z Polski. — Sprawozdania Komisji fizjograficznej Polskiej Akademii Umiejętności **61**: 205–216.

SZEPTYCKI, A. 2007a: Catalogue of the World Protura. — Acta Zool. Cracov., B, **50**: 1–210.

SZEPTYCKI, A. 2007b: Protura. — Fauna Europaea, version 1.3, http://www.faunaeur. org

SZEPTYCKI, A. & CHRISTIAN, E. 2000: Two new *Eosentomon* species from Austria (Insecta: Protura: Eosentomidae). — Ann. Naturhist. Mus. Wien, B, **101**: 83–92.

SZEPTYCKI, A. & CHRISTIAN, E. 2001: *Vindobonella leopoldina* gen. n., sp. n. from Austria (Protura: Acerentomidae s. l.). — Eur. J. Entomol. **98**: 249–255.

THALER, K. 1994: Partielle Inventur der Fauna von Nordtirol: Arachnida, Isopoda: Oniscoidea, Myriapoda, Apterygota (Fragmenta Faunistica Tirolensia – XI). — Ber. nat.-med. Verein Innsbruck **81**: 99–121.

Univ.-Prof. Dr. Erhard CHRISTIAN
Institut für Zoologie, Department für Integrative Biologie
Universität für Bodenkultur
Gregor-Mendel-Straße 33
A-1180 Wien.
E-Mail: erhard.christian@boku.ac.at

Opiliones (Arachnida)

Christian Komposch

Summary. 64 harvestman-species and -subspecies, belonging to 8 families, are currently known from Austria. The most diverse families are Phalangiidae (22 species), followed by Sclerosomatidae and Nemastomatidae (12 each). These mostly carnivorous taxa are, in part, highly stenotopic. Opiliones can be found in very different habitat types in all altitudinal zones. The state of knowledge of Austrian harvestmen is quite good thanks to a 230 year history of exploration. Remarkable is both the decrease in diversity from south to north and the high importance of the Alpine area, which hosts 3 endemic and 10 subendemic harvestman species. The federal state of Carinthia is the most diverse, yielding 54 taxa, in contrast to the Burgenland with just 25 species. The high percentage of endangered species and their value as bioindicators has led to an increasing attention towards harvestmen in modern nature conservation work.

Zusammenfassung: Im Bundesgebiet sind derzeit 64 Weberknechtarten bzw. -unterarten aus insgesamt 8 Familien nachgewiesen. Die artenreichsten Familien sind Phalangiidae (22 Spezies), gefolgt von Sclerosomatidae und Nemastomatidae (je 12). Die vorwiegend räuberisch lebenden und teilweise hoch spezialisierten Arten sind in den unterschiedlichsten Lebensräumen und in allen Höhenstufen zu finden. Die 230-jährige Erforschungsgeschichte der Weberknechte Österreichs führte zu einem guten faunistisch-opilionologischen Durchforschungsgrad des Landes. Hervorzuheben sind das ausgeprägte Süd-Nord-Gefälle der Weberknechtdiversität und die hohe Bedeutung des Alpenraumes, die sich im Auftreten von 3 Endemiten und 10 Subendemiten zeigt. Kärnten ist mit 54 festgestellten Weberknechttaxa das aus sektoraler Sicht artenreichste Bundesland, aus dem Burgenland sind lediglich 25 Spezies bekannt. Als gefährdete Schutzobjekte und hervorragende Bioindikatoren finden Weberknechte im fachlichen Naturschutz zunehmend Beachtung.

Key Words: Opiliones, Arachnida, harvestmen, arachnids, Austria, checklist, biodiversity

Jürgen GRUBER,
dem Großmeister der Weberknechtkunde,
herzlichst gewidmet

I Einleitung

Von den weltweit knapp 6.500 gültig beschriebenen Weberknechtarten (KURY 2010) leben rund zwei Prozent in Mitteleuropa (KOMPOSCH 2006).

Weberknechte oder Kanker (Opiliones) besiedeln fast alle Landlebensräume in hohen Individuendichten. Die Vielfalt an Biotopen reicht dabei von Ruderalfluren über Wald-, Wiesen- und Siedlungslebensräume bis hin zur gletscher- und felsgeprägten Nivalstufe. Überaus beeindruckend ist die hohe Diversität an unterschiedlichsten Erscheinungsformen, die sich in einer Vielfalt an Lebensweisen widerspiegelt. Neben kurzbeinigen, milbenähnlichen und blinden Bodenbewohnern begegnen wir dem "klassischen" Phalangiiden an Felsen und Hausmauern, wobei einige wenige Arten auch die lebensfeindlichen Betonwüsten unserer Städte nicht scheuen. Ein in der Bodenstreu von Schluchtwäldern lebender leuchtend orange gefärbter Cladonychiide ist der einzige heimische Vertreter der Unterordnung Laniatores. Blockhalden- und Höhlen bewohnende Ischyropsalididen faszinieren durch ihre überkörperlangen Cheliceren. Erdummantelte, abgeflachte und für das menschliche Auge kaum auflösbare Troguliden zeigen eine ausgeprägte Thanatose. Die beiden Riesenweberknechte, Gattung *Gyas*, sind anspruchsvolle Bewohner von überhängenden Felswänden und zählen mit einer Spannweite von bis zu 15 Zentimetern zu den größten Arthropoden Europas. Die kryptischen Dicranolasmatiden sind österreichweit auf Arealsplitter im nordöstlichen Österreich beschränkt, der hygrobionte *Paranemastoma bicuspidatum* dringt bis in die aquatische Zone von Quellfluren und Bächen vor.

Danksagung: Die Erstellung der vorliegende Liste wäre ohne die Hilfe zahlreicher Kollegen nicht möglich gewesen. Für das Überlassen von umfangreichem Tiermaterial, Datensätzen und für Diskussion und Hilfe danke ich: Albert AUSOBSKY, Clemens BRANDSTETTER, Jason DUNLOP, Thomas FRIESS, Jürgen GRUBER, Werner HOLZINGER, Andreas KAPP, Ivo KARAMAN, Barbara KNOFLACH-THALER, Brigitte, Harry und Traudi KOMPOSCH, Jochen MARTENS, Christoph MUSTER, Lorenz NEUHÄUSER-HAPPE, Tone und Ljuba NOVAK, Wolfgang PAILL, Günther RASPOTNIG, Axel SCHÖNHOFER, Reinhart SCHUSTER, Konrad THALER (†) und Christian WIESER.

II Allgemeiner Teil

1. Geschichte der Weberknechtforschung in Österreich – Ein kurzer Überblick

Die erste Erwähnung von Opiliones aus Österreich findet sich in SCHRANK (1781). In der ersten Hälfte des 19. Jahrhunderts befasste sich der Wiener Entomologe und Museumskustos Vinzenz KOLLAR mit Weberknechten. Carl Ludwig KOCH war der erste "auswärtige" Zoologe, der grundlegend zur Kenntnis der Weberknechtfauna Österreichs beitrug; viele seiner Namen sind noch heute gültig.

Carl Ludwig DOLESCHALL (s. auch STAGL 1999) publizierte 1852 die erste Artenliste österreichischer Arachniden mit einigen Artbeschreibungen, vielfach gestützt auf von KOLLAR gesammeltes Material und unter Verwendung KOLLAR'scher Manuskriptnamen. Anton AUSSERER (1867) behandelte die Arachniden Tirols (vgl. THALER 1991). Weitere Darstellungen der Fauna Tirols stammen von Camill HELLER (1881a, b) und Ludwig KOCH (1869 ff.). Ein Pionier auf dem Gebiet der Weberknechtbiologie-Forschung war die Innsbrucker Gymnasiallehrerin Hilde STIPPERGER (1928), die vor 80 Jahren ihr Werk "Biologie und Verbreitung der Opilioniden Nordtirols" vorlegte.

Ein unübersehbarer und vielfach verwirrender Einfluss auf die Erforschung der österreichischen Opilionidenfauna ging von dem seinerzeit führenden deutschen Spezialisten Carl Friedrich ROEWER (1923) aus (vgl. KRAUS 1963, von HELVERSEN & MARTENS 1972).

FRANZ (1943, 1949) publizierte in seiner "Landtierwelt der mittleren Hohen Tauern" auch umfangreiche Weberknechtdaten. Gemeinsam mit der Nordostalpen-Monographie (FRANZ & GUNHOLD 1954) sind dies Meilensteine in der faunistischen Erforschung dieser Gebirgszüge, u.a. allerdings durch die zweifelhafte Arbeit und Determination(shilfe) ROEWERS belastet (KOMPOSCH & GRUBER 2004). Dies gilt auch für den Catalogus Faunae Austriae (KRITSCHER 1956), eine recht unkritische Kompilation von Literatur- und Sammlungsdaten.

Von 1958 an bis heute prägte der am Naturhistorischen Museum Wien tätige Weberknechtspezialist Jürgen GRUBER (1960 ff.) die opilionologische Forschungs-landschaft Österreichs wie kein anderer. Daneben war es Konrad THALER (1963 ff.) in Innsbruck, der über Jahrzehnte wertvolle Datensätze zur Weberknechtfauna der Alpen, insbesondere für Nord- und Südtirol, publizierte, später gemeinsam mit sei-nen Studierenden und Mitarbeitern. Hervorzuheben sind hierbei Barbara KNOFLACH, Wilfried BREUSS, Christoph MUSTER, Karl-Heinz STEINBERGER und Vito ZINGERLE.

Albert AUSOBSKY (1987), aufbauend auf Vorarbeiten von Leopold SCHÜLLER (1963), bearbeitete insbesondere die Salzburger Fauna flächendeckend.

Den Meilenstein der Weberknechtforschung Mitteleuropas legte der deutsche Zoologe Jochen MARTENS (1978) mit seinem "Dahl"-Band: weitgehend befreit vom "ROEWER'schen Ballast" (vgl. auch GRUBER 1966) werden Taxonomie, Verbreitung, Ökologie und Biologie der einzelnen Arten erstmals zusammenfassend auf zeit-gemäßem Niveau dargestellt.

Die Steyskal-Kurven (Abb. 1) visualisieren zum einen die Schaffensperioden taxonomisch tätiger Opilionologen, aus österreichischer Sicht hervorzuheben sind C.L. KOCH (12 spp.), J. MARTENS (6), C.F. ROEWER (5), J.F.W. HERBST (4) und J. GRUBER (3), und zeigen zum anderen, dass die Sättigung dieser Kurve noch nicht erreicht ist.

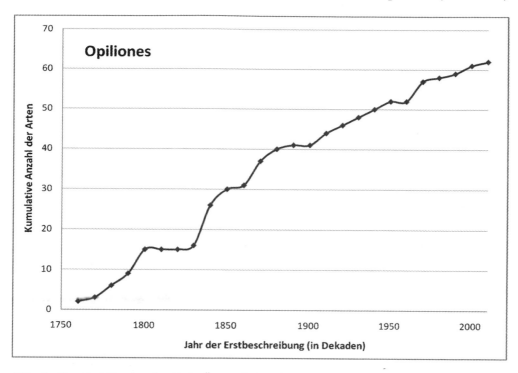

Abb. 1: Steyskal-Kurven für die in Österreich vorkommenden Weberknechtarten und -unter-arten: Jahr der Beschreibung in Dekaden von 1760 bis 2000 (n = 62, für *Leiobunum* sp. und einen Sironiden sind die Daten zum Jahr der Beschreibung noch nicht verfügbar).

Als Sammler in neuerer Zeit wären u.a. Wilhelm KÜHNELT (Bodentiere), Hans STROUHAL (u.a. Katalog der rezenten Höhlentiere Österreichs: STROUHAL & VORNATSCHER 1975) und Reinhart SCHUSTER (Gesiebeproben aus vielen Gebieten Österreichs; Verbreitungskarte für *Cyphophthalmus duricorius*: SCHUSTER 1975), Erich KREISSL (Steiermark), Franz RESSL (Niederösterreich), Alois KOFLER (Osttirol) und Clemens BRANDSTETTER & Andreas KAPP (Vorarlberg) zu erwähnen. In jüngster Zeit widmet sich die Forschergruppe um Günther RASPOTNIG am Institut für Zoologie der Karl-Franzens-Universität Graz den Weberknechten mit chemisch-ökologischen und anatomischen Untersuchungen der Stinkdrüsen und ihren Sekreten. Die weberknechtkundlichen Arbeitsschwerpunkte des Verfassers sind Taxonomie, Faunistik und naturschutzfachliche Aspekte der Opilionenfauna des Alpenraumes.

Einen detaillierten Überblick über die arachnologische und weberknechtkundliche Erforschungsgeschichte Österreichs geben THALER & GRUBER (2003) bzw. KOMPOSCH & GRUBER (2004).

Ch. Komposch

2. Aktueller Forschungsstand

Der faunistisch-opilionologische Durchforschungsgrad von Österreich ist als sehr gut bis gut einzustufen (KOMPOSCH 2002, 2009a; KOMPOSCH & GRUBER 2004; Abb. 2) und manifestiert sich sowohl in der hohen Zahl an verfügbaren Datensätzen als auch an der über die gesamte Landesfläche recht gleichmäßig verteilten zahlreichen Fundorte (für Kärnten siehe KOMPOSCH 1999). Voraussetzung dafür war die Aufklärung der taxonomisch verworrenen und synonymbeladenen Situation, die Carl Friedrich ROEWER (1923 ff.) hinterlassen hatte, insbesondere durch Jürgen GRUBER (Wien) und Jochen MARTENS (Mainz). Seit dem Vorliegen des 64. Bandes der "Tierwelt Deutschlands" (MARTENS 1978) gilt diese Spinnentierordnung mitteleuropaweit als vorbildlich revidiert und gut bestimmbar.

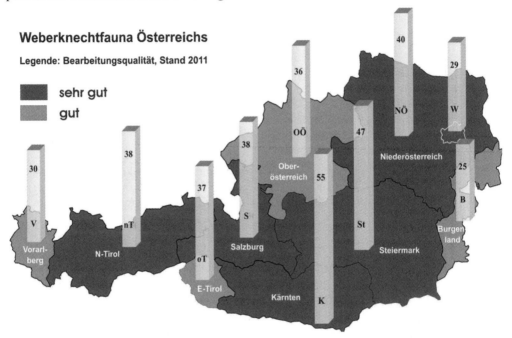

Abb. 2: Bearbeitungsqualität und Weberknecht-Artenzahlen der österreichischen Bundesländer bzw. Landschaftsteile. [nach KOMPOSCH 2009a]

Das hohe wissenschaftliche und naturschutzfachlich-angewandte Interesse an Opilioniden in Europa zeigt sich in einer Vielzahl an publizierten Checklisten und Roten Listen in den letzten Jahren. Eine Übersicht für Mittel- und Nordeuropa geben BLICK & KOMPOSCH (2004).

Die Weberknechtfauna Österreichs zeigt sich mit den 64 nachgewiesenen Taxa im sektoralen Biodiversitätsvergleich mit den Nachbarländern vergleichsweise artenreich (Schweiz: 50 spp., Deutschland: 52 spp., Tschechische Republik: 34 spp., Slowakei: 33 spp., Ungarn: 35 spp., Slowenien: 66 spp., Italien: 124 spp., davon 33 spp. in Südtirol; HELLRIGL 1996; MURÁNYI 2005; NOVAK & GRUBER 2000, NOVAK et al. 2006; G. LENGYEL in litt.; Übersichtsdarstellungen liefern BLICK & KOMPOSCH 2004 und KOMPOSCH & GRUBER 2004; ergänzt durch SCHÖNHOFER & HOLLE 2007, WIJNHOVEN et al. 2007; NOVAK 2005 ergänzt CHEMINI 1994; NOVAK & GIRIBET 2006; vgl. KOMPOSCH 2009a).

3. Methoden und Datengrundlage

Die Benennung der Arten erfolgt im Allgemeinen nach MARTENS (1978), die Familienzuordnung nach KURY (2003). Die Familienzuordnung und -benennung unterscheidet sich von MARTENS (1978) zum einen in der Übernahme des Vorschlags COKENDOLPHERS (1985) Erebomastridae durch Cladonychiidae zu ersetzen und zum anderen in der Abtrennung der Sclerosomatidae von den Phalangiidae (sensu MARTENS 1978) auf Familienniveau (vgl. CRAWFORD 1992, STAREGA 2000, KURY 2003, TOURINHO 2007) trotz unwahrscheinlicher Monophylie der Phalangiidae in dieser Zusammensetzung (vgl. BLICK & KOMPOSCH 2004, KOMPOSCH 2009a, J. GRUBER in litt.).

Von den Synonymen werden vor allem jene genannt, die seit dem Erscheinen des Weberknechtbandes im Rahmen der Tierwelt Deutschlands (MARTENS 1878) bekannt wurden.

Für aus Österreich beschriebene, aktuell gültige Arten wird der locus typicus genannt, nicht jedoch für Synonyme.

Als Datengrundlagen standen neben publizierten auch zahlreiche unpublizierte Arbeiten und Sammlungsprotokolle mit weberknechtkundlichen Inhalten zur Verfügung. Die Anzahl der exakt verorteten und digitalisierten opionologischen Datensätze beträgt derzeit mehr als 13.500 (Datenbank Ch. KOMPOSCH/ÖKOTEAM; Stand Juni 2010; vgl. KOMPOSCH 2009a).

4. Lebensweise und Gefährdungsstatus

Der Großteil der Weberknechtarten ernährt sich räuberisch. Erbeutet werden Insekten, Spinnentiere, Krebstiere, Würmer, Hundertfüßer und Schnecken, oft verletzte oder frisch verendete Tiere (u.a. KÄSTNER 1926). Umgekehrt ist die Zahl der Fressfeinde für Weberknechte sehr hoch.

Die Fortpflanzung der Weberknechte erfolgt im Allgemeinen durch eine echte Kopula. Die Eiablage erfolgt in Hohlräume des Bodens, von Pflanzen oder in Totholz, bei Troguliden in leere Schneckenhäuser (u.a. PABST 1953). Vielfach überwintern Weberknechte im Ei- oder Juvenilstadium. Nach vier bis acht Reifehäutungen erreichen die Tiere ihre Geschlechtsreife. Evolutionsbiologisch "moderne" Weberknechte (Phalangiidae, Sclerosomatidae) erreichen meist ein Alter von knapp einem Jahr, ursprünglichere Bodenbewohner leben einige wenige Jahre; das Höchstalter wurde bei den Cyphophthalmi mit neun Jahren festgestellt.

Ch. Komposch

Von den 64 für Österreich belegten Weberknechtarten gelten aktuell 12 Arten (18,8 %) als "Gefährdet" (VU), 17 (26,6 %) als "Stark gefährdet" (EN) und 6 (9,4 %) als "Vom Aussterben bedroht" (CR) (KOMPOSCH 2009a). Für die beiden aktuell nachgewiesenen Cyphophthalmi liegen noch keine Gefährdungseinstufungen vor. Im fachlichen Naturschutz gewinnen Weberknechte vermehrt Beachtung als hervorragende Biotopdeskriptoren und Bioindikatoren.

5. Taxonomischer und geographischer Überblick

Die vorliegende Checkliste enthält insgesamt 64 für Österreich bekannte Weberknechtarten (Stand: März 2011).

Damit ist das Weberknecht-Artenspektrum für das Bundesgebiet als weitgehend erfasst anzusehen. Zu erwarten sind noch versteckte Arten innerhalb der revisionsbedürftigen Gattung *Trogulus*. Daneben könnten noch weitere Spezies mit grenznahen Vorkommen im benachbarten Ausland und weitere Neozoen die Fauna Austriaca bereichern (KOMPOSCH 1997; KOMPOSCH & GRUBER 2004).

Die Verteilung der Weberknechtarten Österreichs auf die einzelnen Familien ist Tabelle 1 zu entnehmen.

Der Durchschnittswert von nur 8 Weberknechtarten pro Familie zeigt die relativ hohe Diversität an in Österreich vertretenen Familien.

Tab. 1: Verteilung der in Österreich festgestellten 64 Weberknechttaxa auf Familien. Die Artenzahl der Trogulidae (8) dürfte sich nach Abschluss der laufenden Revisionsarbeiten auch im Bundesgebiet erhöhen.

Familie	Artenzahl	%
Phalangiidae	22	34,4
Sclerosomatidae	12	18,7
Nemastomatidae	12	18,7
Trogulidae	8	12,5
Ischyropsalididae	5	7,8
Sironidae	3	4,7
Cladonychiidae	1	1,6
Dicranolasmatidae	1	1,6
Total	**64**	**100**

Bemerkenswert sind das ausgeprägte Süd-Nord-Gefälle bezüglich der Weberknechtdiversität (Kärnten: 55 spp., Steiermark: 47 und Niederösterreich: 40) und die hohe Bedeutung des Alpenraumes für diese Spinnentierordnung, welche sich im Vorhandensein von 3 Endemiten und 10 Subendemiten zeigt.

6. Anmerkungen zu den Artkommentaren

Angaben zur Horizontalverbreitung in Österreich werden hinsichtlich des Bundeslandvorkommens geliefert. Aus naturräumlichen und zoogeographischen Gründen wird – dem Catalogus Faunae Austriae (KRITSCHER 1956) – folgend zwischen Osttirol und Nordtirol unterschieden, im Gegensatz zu jenem Werk wird jedoch, da aus landesfaunistischer Sicht bedeutsam, die Trennung zwischen Wien und Niederösterreich aufrechterhalten.

Kürzel der Bundesländer (bzw. Landschaftsteile):

B	=	Burgenland
K	=	Kärnten
N	=	Niederösterreich
nT	=	Nordtirol
O	=	Oberösterreich
oT	=	Osttirol
S	=	Salzburg
St	=	Steiermark
V	=	Vorarlberg
W	=	Wien

Nennungen von Synonymen beschränken sich auf ausgewählte, in MARTENS (1978) meist nicht enthaltene Namen. Für sämtliche aus Österreich beschriebenen validen Arten wird der locus typicus angeführt.

Endemiten und Subendemiten Österreichs sind als solche ausgewiesen. Unter Subendemit werden an dieser Stelle jene Taxa verstanden, deren Areal zumindest zu 75 % in Österreich liegt (KOMPOSCH 2009b).

III Spezieller Teil

1. Liste der in Österreich vorkommenden Weberknechtarten

Familie Sironidae

Cyphophthalmus duricorius (JOSEPH, 1868)
Synonym: *Siro duricorius* (JOSEPH, 1868)
Verb.: K, St

Ch. Komposch

Siro cf. *crassus* Novak & Giribet, 2006
>   Verbr.: St
>   Subendemit

Sironidae gen. et sp. nov.?
>   Synonym: *Siro* nov. sp.? (in Komposch & Gruber 2004: 524)
>   Verbr. K, St
>   Endemit?

Familie Cladonychiidae

Holoscotolemon unicolor Roewer, 1915
>   Verb.: K, N, nT, O, oT, S, St
>   locus typicus: Lienz, oT
>   Subendemit

Familie Nemastomatidae

Carinostoma carinatum (Roewer, 1914)
>   Synonym: *Mitostoma carinatum* (Roewer, 1914)
>   Verb.: K, oT

Histricostoma dentipalpe (Ausserer, 1867)
>   Verb.: K, nT, oT, S, V
>   locus typicus: Umgebung Innsbruck, Husslhof, nT

Mitostoma alpinum (Hadži, 1931)
>   Verb.: K, N, St
>   Subendemit
>   in den Südl. und Nördl. Kalkalpen, hier ev. mit eigenständiger Art vertreten

Mitostoma chrysomelas (Hermann, 1804)
>   Verb.: B, K, N, nT, O, oT, S, St, V, W

Nemastoma bidentatum bidentatum Roewer, 1914
>   Verb.: K, St
>   locus typicus: Karawanken, Feistritz, K

Nemastoma bidentatum relictum Gruber & Martens, 1968
>   Verb.: K, S, St
>   locus typicus: Kleinarltal, Tappenkarsee, S
>   Endemit
>   Artstatus?

Nemastoma bidentatum sparsum GRUBER & MARTENS, 1968
    Verb.: B, K, N, St, W
    locus typicus: Leithagebirge, Sonnenberg, B

Nemastoma lugubre (MÜLLER, 1776)
    Verb.: N, O, V, W

Nemastoma schuelleri GRUBER & MARTENS, 1968
    Verb.: K, nT, S, St
    locus typicus: Rotgülden, WNW Muhr, Murtal, Lungau, S
    Endemit

Nemastoma triste (C. L. KOCH, 1835)
    Verb.: K, N, nT, O, oT, S, St, V, W
    postglazial reliktäre Form mit Hauptverbreitungsgebiet in den Alpen

Paranemastoma bicuspidatum (C. L. KOCH, 1835)
    Verb.: K, nT, O, oT, S, St
    locus typicus: Hohe Tauern, Gastein, S
    Subendemit

Paranemastoma quadripunctatum (PERTY, 1833)
    Verb.: B, K, N, nT, O, oT, S, St, V, W

Familie Dicranolasmatidae

Dicranolasma scabrum (HERBST, 1799)
    Verb.: B, N
    großräumig isolierter Vorposten im östlichsten Bundesgebiet

Familie Trogulidae

Anelasmocephalus cambridgei (WESTWOOD, 1874)
    Verb.: V
    nur punktuelle Vorkommen dieser atlantisch-submediterran verbreiteten Art
    in der Rheinebene Vorarlbergs

Anelasmocephalus hadzii MARTENS, 1978
    Verb.: K, N, O, oT, St
    locus typicus: Karawanken, Eisenkappel, K

Ch. Komposch

Trogulus cisalpinus CHEMINI & MARTENS, 1988
    Verb.: K
    bislang nur vom Kärntner Anteil der Steiner Alpen und aus den Karnischen
    Alpen bekannt

Trogulus closanicus AVRAM, 1971
    Verb.: B, K, N, nT, O, St, W

Trogulus falcipenis KOMPOSCH, 2000
    Verb.: K
    locus typicus: Karawanken, Koschuta, K

Trogulus nepaeformis (SCOPOLI, 1763)
    Verb.: B, K, N, nT, O, oT, S, St, V, W

Trogulus tingiformis C. L. KOCH, 1848
    Verb.: K, N, O, S, St, W

Trogulus tricarinatus (LINNAEUS, 1767)
    Verb.: B, K, N, nT, O, oT, S, St, V, W

Familie Ischyropsalididae

Ischyropsalis carli LESSERT, 1905
    Verb.: nT, V

Ischyropsalis dentipalpis CANESTRINI, 1872
    Synonym: *Ischyropsalis helvetica* ROEWER, 1916
    Verbr.: nT, V

Ischyropsalis hadzii ROEWER, 1950
    Verb.: K
    Subendemit
    einziger troglobionter Weberknecht Österreichs

Ischyropsalis hellwigii hellwigii (PANZER, 1794)
    Synonym: *Ischyropsalis hellwigi hellwigi* (PANZER, 1794)
    Verb.: K, N, nT, O, S, St, V, W

Ischyropsalis kollari C. L. KOCH, 1839
    Verb.: K, N, nT, O, oT, S, St
    locus typicus: Hohe Tauern, Gastein, S
    Subendemit

Familie Phalangiidae

Amilenus aurantiacus (SIMON, 1881)
    Verb.: B, K, N, nT, O, oT, S, St, V, W
    Überwinterung in Stollen und Höhlen, Massenaggregationen

Dasylobus graniferus (CANESTRINI, 1871)
    Synonyme: *Eudasylobus nicaeensis* (THORELL, 1876)
    E. roeweri STIPPERGER, 1928
    Verb.: nT, V
    Forschungsdefizite bestehen bezüglich dieser wenig bekannten Art in
    Österreich

Dicranopalpus gasteinensis DOLESCHALL, 1852
    Verb.: K, N, nT, O, oT, S, St, V
    locus typicus: Hohe Tauern, Gastein, S

Egaenus convexus (C. L. KOCH, 1835)
    Verb.: B, K, N, St, W
    locus typicus: Umgebung Wien, W/N

Lacinius dentiger (C. L. KOCH, 1848)
    Synonyme: *Odiellus remyi* (DOLESCHALL, 1852), *O. remyi* ROEWER, 1923
    Verb.: B, K, N, nT, O, oT, S, St, V, W
    locus typicus: Salzburg, S

Lacinius ephippiatus (C. L. KOCH, 1835)
    Verb.: B, K, N, nT, O, oT, S, St, V, W
    locus typicus: Hohe Tauern, Gastein, S

Lacinius horridus (PANZER, 1794)
    Verb.: B, K, N, O, S, St, W

Lophopilio palpinalis (HERBST, 1799)
    Verb.: B, K, N, nT, O, oT, S, St, V, W

Megabunus armatus (KULCZYNSKI, 1887)
    Verb.: K, oT
    Subendemit

Megabunus lesserti SCHENKEL, 1927
    Verb.: K, N, nT, O, S, St
    Subendemit

Mitopus glacialis (HEER, 1845)
    Synonym: *Parodiellus obliquus* (C. L. KOCH, 1839)

Ch. Komposch

Verb.: K, nT, oT, S, St, V
kaltstenotherme Art der Alpin- und Nivalstufe ("Gletscherweberknecht")

Mitopus morio (FABRICIUS, 1779)
Synonym: *Mitopus morio* (FABRICIUS, 1799)
Verb.: B, K, N, nT, O, oT, S, St, V, W

Oligolophus tridens (C. L. KOCH, 1836)
Verb.: B, K, N, nT, O, oT, S, St, V, W

Opilio canestrinii (THORELL, 1876)
Synonym: *Opilio ravennae* SPOEK, 1962 (ad part.)
Verb.: B, K, N, nT, O, oT, S, St, V, W
wichtigstes Neozoon, dominiert inzwischen die synanthropen
Weberknechtzönosen

Opilio dinaricus ŠILHAVÝ, 1938
Verb.: K, N, O, oT, S, St

Opilio parietinus (DE GEER, 1778)
Verb.: B, K, N, nT, O, oT, S, St, W

Opilio ruzickai ŠILHAVÝ, 1938
Verb.: B, K, N, St, W
Neozoon

Opilio saxatilis C. L. KOCH, 1839
Verb.: B, K, N, nT, O, oT, S, St, V, W

Phalangium opilio LINNAEUS, 1758
Synonym: *Phalangium opilio* LINNAEUS, 1761
Verb.: B, K, N, nT, O, oT, S, St, V, W

Platybunus bucephalus (C. L. KOCH, 1835)
Verb.: K, N, nT, O, oT, S, St

Platybunus pinetorum (C. L. KOCH, 1839)
Verb.: K, N, nT, O, oT, S, St, V

Rilaena triangularis (HERBST, 1799)
Verb.: B, K, N, nT, O, oT, S, St, V, W

Familie Sclerosomatidae

Astrobunus helleri (AUSSERER, 1867)
Synonyme: ?*Astrobunus glockneri* ROEWER, 1957,

?*A. bavaricus* ROEWER, 1957 (siehe KOMPOSCH & GRUBER 2004)
Verb.: K, nT, oT
locus typicus: Umgebung Innsbruck, Husslhof, nT

Astrobunus laevipes (CANESTRINI, 1872)
Verb.: B, K, N, O, St, W

Gyas annulatus (OLIVIER, 1791)
Verb.: K, nT, oT, S

Gyas titanus SIMON, 1879
Verb.: K, N, nT, O, S, St, V

Leiobunum limbatum L. KOCH, 1861
Verb.: K, N, nT, O, oT, S, St, V, W

Leiobunum roseum C. L. KOCH, 1839
Synonym: *Leiobunum purpurissatum* L. KOCH, 1869
Verb.: K, oT
Subendemit

Leiobunum rotundum (LATREILLE, 1798)
Verb.: B, K, N, nT, O, oT, S, St, V, W

Leiobunum rupestre (HERBST, 1799)
Verb.: B, K, N, nT, O, oT, S, St, V, W

Leiobunum subalpinum KOMPOSCH, 1998
Verb.: K, S, St, oT
locus typicus: Hohe Tauern, Gößnitztal, K
Subendemit

Leiobunum sp.
Verb.: V
invasives Neozoon unbekannter Herkunft, Artzugehörigkeit derzeit unklar,
Massenvorkommen an anthropogenen Strukturen (Details siehe WIJNHOVEN
et al. 2007)

Nelima apenninica MARTENS, 1969
Verb.: oT
ein rezenter Wiederfund dieser verschollen geglaubten Art in Osttirol

Nelima sempronii SZALAY, 1951
Synonym: *Nelima semproni* SZALAY, 1951
Verb.: B, K, N, nT, O, oT, S, St, W

Ch. Komposch

2. Fehlmeldungen

Der Catalogus Faunae Austriae (KRITSCHER 1956) listet eine Reihe fragwürdiger Namen auf – teils "alte" Fehldeutungen und Dubiosa, teils vor allem auf ROEWER zurückgehende überflüssige Namen (jüngere Synonyme) und "Fantasiefundorte", auch von HADŽI's Wiederbeschreibungen alter, aber von ROEWER ungenügend dargestellter Arten herstammende Mehrfachbenennungen! (KOMPOSCH & GRUBER 2004: 525). Insgesamt werden im alten Catalogus 80 Arten angeführt, von denen lediglich 35 (44 %) valide sind!

Die im Folgenden aufgelisteten Fehlmeldungen werden von KOMPOSCH & GRUBER (2004) kommentiert, bezüglich weiterer Fehlmeldungen, Synonymien und Fundortverwechslungen siehe GRUBER (1964, 1966 ff.), HELVERSEN & MARTENS (1972), MARTENS (1969, 1978) und THALER & GRUBER (2003):

Astrobunus glockneri ROEWER, 1957
Ischyropsalis janetscheki ROEWER, 1950
Ischyropsalis pestae ROEWER, 1950
Nemastoma janetscheki SCHENKEL, 1950
Nemastoma riparium ROEWER, 1951
Nemastoma schenkeli ROEWER, 1951
Nemastoma sillii HERMANN, 1871
Nemastoma werneri KULCZYŃSKI, 1903
Platybunus exiguus ROEWER, 1956

IV Literatur

AUSOBSKY, A. 1987: Verbreitung und Ökologie der Weberknechte (Opiliones, Arachnida) des Bundeslandes Salzburg. — Jahrbuch Haus der Natur **10**: 40–52.

AUSSERER, A. 1867: Die Arachniden Tirols nach ihrer horizontalen und verticalen Verbreitung. — Verh. k.k. Zool.-Bot. Ges.Wien **17**: 137–170.

BLICK, T. & KOMPOSCH, Ch. 2004: Checkliste der Weberknechte Mittel- und Westeuropas. / Checklist of the harvestmen of Central and Western Europe (Arachnida: Opiliones). — Internet: http://www.arages.de/files/checklist2004_opiliones.pdf. 6 pp.

CHEMINI, C. 1994: Arachnida. Scorpiones, Palpigradi, Solifugae, Opiliones. — In MINELLI, A., RUFFO, S. & POSTA, S. la (eds.): Checklist delle specie della fauna italiana **21**: 1–42. — Calderini, Bologna.

COKENDOLPHER, J.C. 1985: Erebomastridae: replaced by Cladonychiidae (Arachnida: Opiliones). — Entomological News **96**: 36.

CRAWFORD, R. L. 1992: Catalogue of the genera and type species of the harvestman superfamily Phalangioidea (Arachnida). — Burke Museum Contributions in Anthropology and Natural History **8**: 1–60.

FRANZ, H. 1943: Die Landtierwelt der mittleren Hohen Tauern. Ein Beitrag zur Tiergeographischen und -soziologischen Erforschung der Alpen. — Denkschr. Akad. Wiss. Wien, math.- nat. Kl. **107**: 552 pp. + 14 Tafeln u. 10 Karten.

FRANZ, H. 1949: Erster Nachtrag zur Landtierwelt der mittleren Hohen Tauern. — Sitzungsb. österr. Akad. Wiss. Wien **158 A 1** (1–2): 1–77.

FRANZ, H. & GUNHOLD, P. 1954: 19. Ordnung Opiliones. — In FRANZ, H.: Die Nordostalpen im Spiegel ihrer Landtierwelt. Eine Gebietsmonographie Bd. **1**, pp. 461–472. — Innsbruck: Universitätsverlag Wagner.

GRUBER, J. 1960: Ein Beitrag zur Kenntnis der Opilionenfauna des Leithagebirges und der Hainburger Berge. — Bgld. Heimatblätter **22**: 117–126.

GRUBER, J. 1964: Kritische und ergänzende Beobachtungen zur Opilionidenfauna Österreichs (Arachnida). — Z. Arbeitsgem. Österr. Entomol. **16**: 1–5.

GRUBER, J. 1966: Neues zur österreichischen Opilionidenfauna (Arachnida). — Z. Arbeitsgem. Österr. Entomol. **18**: 43–47.

HELLER, C. 1881a: Über die Verbreitung der Thierwelt im Tiroler Hochgebirge. I. Abtheilung. — Sitzungsb. österr. Akad. Wiss., Math.-Nat. Cl. **83** (I): 103–175.

HELLER, C. 1881b: Über die Verbreitung der Thierwelt im Tiroler Hochgebirge. II. Abtheilung. — Sitzungsb. österr. Akad. Wiss., Math.-Nat. Cl. **86** (I): 8–53.

HELLRIGL, K. 1996: Opiliones – Weberknechte, Afterspinnen. — In HELLRIGL, K. (Hrsg.): Die Tierwelt Südtirols. — Veröffentlichungen des Naturmuseums Südtirol, Bozen **1**: 205–210.

HELVERSEN, O. von & MARTENS, J. 1972: Unrichtige Fundort-Angaben in der Arachniden-Sammlung ROEWER. — Senckenbergiana biol. **53**: 109–123.

KÄSTNER, A. 1926: Opiliones. Weberknechte. — In SCHULZE, P. (Hrsg.): Biologie der Tiere Deutschlands **19**: 1–55. — Berlin: Gebrüder Borntraeger.

KÄSTNER, A. 1928: Opiliones (Weberknechte, Kanker). — In DAHL, F. (Hrsg.): Die Tierwelt Deutschlands und der angrenzenden Meeresteile nach ihren Merkmalen und nach ihrer Lebensweise. 8. Teil. Spinnentiere oder Arachnoidea. III: Opiliones – Pseudoscorpionida – Pantopoda – Pentastomida, pp. 1–51. — Jena: Gustav-Fischer.

KOCH, L. 1869-1872: Beitrag zur Kenntnis der Arachnidenfauna Tirols. — Zeitschrift Ferdinandeum Innsbruck **14** (1869): 149–206; **17** (1872): 239–328.

KOMPOSCH, Ch. 1997: Kommentierte Checkliste der Weberknechte (Opiliones) Kärntens. — Carinthia II **187./107.**: 597–608.

KOMPOSCH, Ch. 1999: Rote Liste der Weberknechte Kärntens (Arachnida: Opiliones). — Naturschutz in Kärnten **15**: 547–565.

KOMPOSCH, Ch. 2002: Taxonomie, Faunistik und Ökologie südostalpiner Weberknechte (Arachnida, Opiliones). — Entomologica Austriaca **6**: 19–20.

KOMPOSCH, Ch. 2006: Weberknechte. — In BROCKHAUS-REDAKTION (Hrsg.): Brockhaus-Enzyklopädie. Faszination Natur. Tiere, Bd. **2** (Wirbellose II), pp. 44–47. — Leipzig: F. A. Brockhaus.

KOMPOSCH, Ch. 2009a: Rote Liste der Weberknechte Österreichs (Arachnida: Opiliones). — In ZULKA, P. (Red.): Rote Listen gefährdeter Tiere Österreichs.

Checklisten, Gefährdungsanalysen, Handlungsbedarf. — Grüne Reihe des Lebensministeriums **14/3**: 397–483.

KOMPOSCH, Ch. 2009b: Weberknechte (Opiliones). — In RABITSCH, W. & ESSL, F. (Red.): Endemiten – Kostbarkeiten in Österreichs Pflanzen- und Tierwelt, pp. 476–496. — Naturwissenschaftlicher Verein für Kärnten, Klagenfurt und Umweltbundesamt, Wien.

KOMPOSCH, Ch. & GRUBER, J. 2004: Die Weberknechte Österreichs (Arachnida: Opiliones). — Denisia **12**, zugleich Kataloge der OÖ. Landesmuseen Neue Serie **14**: 485–534.

KRAUS, O. 1963: Carl-Friedrich ROEWER 1881–1963. — Senckenbergiana biol. **44**: 553–562.

KRITSCHER, E. 1956: Opiliones. — Catalogus Faunae Austriae **9c**: 1–8.

KURY, A. B. 2003: Checklist of valid genera of Opiliones of the world. — Internet: http://acd.ufrj.br/mndi/Aracnologia/checklaniator.htm

KURY, A. B. 2010: Classification of Opiliones. Museu Nacional/UFRJ website. — Internet: http://www.museunacional.ufrj.br/mndi/Aracnologia/opiliones.html

MARTENS, J. 1969: Die Abgrenzung von Biospezies auf biologisch-ethologischer und morphologischer Grundlage am Beispiel der Gattung *Ischyropsalis* C.L. KOCH, 1839 (Opiliones, Ischyropsalididae). — Zool. Jb., Abt. System. **96**: 133–264.

MARTENS, J. 1978: Spinnentiere, Arachnida: Weberknechte, Opiliones. — In SENGLAUB, F., HANNEMANN, H.J. & SCHUMANN, H. (Eds.): Die Tierwelt Deutschlands **64**: 464 pp. — Jena: Gustav Fischer.

MURÁNYI, D. 2005: *Amilenus aurantiacus* (SIMON, 1881) (Opiliones), new to Hungary. — Folia Entomologica Hungarica **66**: 7–8.

NOVAK, T. 2005: Harvestmen of the museo Friulano di storia naturale in Udine (Arachnida: Opiliones). Part I. Gortania — Atti del Museo Friulano di Storia Naturale **26**: 211–41.

NOVAK, T. & GIRIBET, G. 2006: A new species of Cyphophthalmi (Arachnida, Opiliones, Sironidae) from Eastern Slovenia. – Zootaxa **1330**: 27-42.

NOVAK, T. & GRUBER, J. 2000: Remarks on published data on harvestmen (Arachnida: Opiliones) from Slovenia. — Annales, Ser. hist. nat. **10** (2000): 281–308.

NOVAK, T., LIPOVŠEK DELAKORDA, S. & SLANA NOVAK, L. 2006: A review of harvestmen (Arachnida: Opiliones) in Slovenia. — Zootaxa **1325**: 267–276.

PABST, W. 1953: Zur Biologie der mitteleuropäischen Troguliden. — Zool. Jb., Abt. Syst., Ökol. u. Geogr. d. Tiere **82**: 1–156.

RASPOTNIG, G., GRUBER, J., KOMPOSCH, Ch., SCHUSTER, R., FÖTTINGER, P., SCHWAB, J. & KARAMAN, I. 2011: Wie viele Arten von Milbenkankern (Opiliones, Cyphophthalmi) gibt es in Österreich? – Arachnologische Mitteilungen, **41** (in prep).

ROEWER, C.F. 1923: Die Weberknechte der Erde. Systematische Bearbeitung der bisher bekannten Opiliones, 1116 S. — Jena: Gustav Fischer.

SCHÖNHOFER, A.L. & HOLLE, T. 2007: *Nemastoma bidentatum* (Arachnida: Opiliones): neu für Deutschland und die Tschechische Republik. — Arachnologische Mitteilungen **33**: 25–30.

SCHÖNHOFER, A.L. & MARTENS, J. 2010: On the identity of *Ischyropsalis dentipalpis* CANESTRINI, 1872 and the description of *Ischyropsalis lithoclasica* sp. n. (Opiliones: Ischyropsalididae). — Zootaxa **2613**: 1–14.

SCHRANK, F. de Paula 1781: Enumeratio Insectorum Austria Indigenorum. — Augustae Vindelicorum, Vid. Eberh. Klett et Franck, 548 pp.

SCHÜLLER, L. 1963: Die Weberknechte des Landes Salzburg. pp. 134–138. — In NATURWISSENSCHAFTLICHE ARBEITSGEMEINSCHAFT AM HAUS DER NATUR IN SALZBURG (Hrsg.): Die naturwissenschaftliche Erforschung des Landes Salzburg. Stand 1963. Gewidmet Herrn Prof. Eduard Paul TRATZ anläßlich seines 75. Geburtstages.

SCHUSTER, R. 1975: Die Verbreitung des Zwergweberknechtes *Siro duricorius* (JOSEPH) in Kärnten [Opiliones, Cyphophthalmi]. — Carinthia II **165./85.**: 285–289.

STAGL, V. 1999: Carl Ludwig DOLESCHALL – Arzt, Forscher und Sammler. — Quadrifina **2**: 195–203.

STARĘGA, W. 2000: Check-list of harvestmen (Opiliones) of Poland. — Internet: http://www.arachnologia.edu.pl/kosarze.html

STIPPERGER, H. 1928: Biologie und Verbreitung der Opilioniden Nordtirols. — Arbeiten aus dem Zoologischen Institut der Universität Innsbruck **3**: 19–79.

THALER, K. 1963: Spinnentiere aus Lunz (Niederösterreich) nebst Bemerkungen zu einigen von KULCZYNSKI aus Niederösterreich gemeldeten Arten. — Ber. Nat.-Med. Ver. Innsbruck **53** (1959-63, Festschrift H. GAMS): 273–283.

THALER, K. 1991: Beiträge zur Spinnenfauna von Nordtirol – 1. Revidierende Diskussion der "Arachniden Tirols" (Anton AUSSERER 1867) und Schrifttum. — Veröffentlichungen des Tiroler Landesmuseums Ferdinandeum (Innsbruck) **71**: 155–189.

THALER, K. & GRUBER, J. 2003: Zur Geschichte der Arachnologie in Österreich 1758–1955. — Denisia **8**: 139–163.

TOURINHO, A. L. 2007: Sclerosomatidae SIMON, 1879. — In PINTO-DA-ROCHA, R., MACHADO, G. & GIRIBET, G. (Eds.): Harvestmen. The biology of Opiliones, pp. 128–131. — Cambridge: Harvard Univ. Press.

WIJNHOVEN, H., SCHÖNHOFER, A. L. & MARTENS, J. 2007: An unidentified harvestman *Leiobunum* sp. alarmingly invading Europe (Arachnida: Opiliones). — Arachnologische Mitteilungen **34**: 27–38.

Mag. Dr. Christian KOMPOSCH
ÖKOTEAM – Institut für Tierökologie und Naturraumplanung
Bergmanngasse 22, A-8010 Graz, Österreich
E-mail: c.komposch@oekoteam.at
Internet: www.oekoteam.at

Pseudoscorpiones (Arachnida)

Volker MAHNERT

Summary: Ten families and 71 species of pseudoscorpions are recorded from Austria, two of them for the first time (*Chthonius (E.) microtuberculatus* HADŽI, *Roncus tenuis* HADŽI). Seven species can be considered as endemic taxa.

Zusammenfassung: Zehn Familien und 71 Arten von Pseudoskorpionen sind aus Österreich gemeldet, darunter Erstmeldungen für zwei Arten (*Chthonius (E.) micro-tuberculatus* HADŽI und *Roncus tenuis* HADŽI). Sieben Arten werden als Endemiten eingestuft.

Key Words: Arachnida, Pseudoscorpiones, Austria, checklist, biodiversity

I Einleitung

Pseudoskorpione bilden eine relativ kleine, gut definierte Arachniden-Ordnung, die den Solifugae (Walzenspinnen) phylogenetisch nahe steht. Eine moderne phylogenetische Analyse und Klassifizierung der Pseudoskorpione wurde von HARVEY (1992) vorgelegt. Nach HARVEY (2009) sind derzeit 26 Familien mit mehr als 3400 gültig betrachteten Arten in 442 Gattungen weltweit bekannt. In Österreich sind sie von den Tallagen bis in die Alpinstufe verbreitet; sie sind feuchtigkeitsliebende Streu- und Humusbewohner (Chthoniidae, Neobisiidae, Syarinidae), andere wiederum sind oft unter Baumborke zu finden (Chernetidae, Cheliferidae), manche andere besiedeln Nester von Insekten, Vögeln oder Säugetieren. Synanthrop finden sich regelmäßig *Cheiridium museorum* und *Chelifer cancroides*. Die erste Meldung eines Pseudoskorpions dürfte auf Aristoteles (384-322 v. Chr.) zurückgehen, der ein "Thierchen mit Scheren, das sich in Büchern aufhält" zitiert (nach LENZ, 1856/1966, S. 531, *Chelifer cancroides*).

II Allgemeiner Teil

1. Erforschungsgeschichte und aktueller Forschungsstand in Österreich

Erste Angaben für das heutige Bundesgebiet finden sich bei L. KOCH (1873, 1876). Es wird auf die Zusammenfassungen von MAHNERT (2004, 2009) hingewiesen, doch sei unterstrichen, dass unser Wissen über Pseudoskorpione untrennbar vereint ist mit dem jahrzehntenlangen Werk von Max BEIER, der am Naturhistorischem Museum Wien tätig war. Zitiert seien hier nur drei zusammenfassende Veröffentlichungen: der Catalogus Faunae Austriae (BEIER 1952, 1956) und der Band zur Bodenfauna Europas (BEIER 1963). Großen faunistischen und ökologischen Wissenszuwachs verdanken wir der intensiven Tätigkeit von Franz RESSL, Purgstall (siehe RESSL 1983, RESSL & BEIER 1958). Andererseits leisten "kleine" z.T. kaum bekannte Veröffentlichungen wertvolle faunistische Beiträge (z.B. KOFLER 1972 oder SCHÜLLER 1951). Erwähnt muss auch die Checkliste der Pseudoskorpione Mitteleuropas (BLICK et al. 2004) werden, die sich für österreichische Daten auf die Zusammenstellung durch MAHNERT (2004) stützt. In dieser Veröffentlichung finden sich zahlreiche faunistisch wichtige Literaturdaten.

III Spezieller Teil

1. Liste der in Österreich nachgewiesenen Arten und Unterarten

Abkürzungen der Bundesländer

B	=	Burgenland
K	=	Kärnten
N	=	Niederösterreich (inkl. Wien)
O	=	Oberösterreich
S	=	Salzburg
St	=	Steiermark
T	=	Tirol
V	=	Vorarlberg
W	=	Wien

(bei älteren Fundangaben wird Wien zu Niederösterreich gerechnet)

Bei aus Österreich beschriebenen Arten ist der locus typicus (loc. typ.) angeführt. Mit Sternchen* werden Erstnachweise für das betreffende Bundesland oder für Österreich (unpubl., det. V. MAHNERT) hervorgehoben. Mit "Ö" (= Österreich) wird die Verteilung einiger weniger Arten charakterisiert, wobei ich mich auf BEIER (1952) und seine z.T. unpublizierten Daten stütze; so ist *Allochernes powelli* "in allen Bundesländern verbreitet" (BEIER in litt., in KOFLER 1972), jedoch publiziert nur für Niederösterreich und Tirol. Synonymien sind bei HARVEY (2009) auffindbar. Die Familien werden in alphabetischer Anordnung angeführt, hiermit HARVEY (2009) folgend. Das Literaturverzeichnis beinhaltet Veröffentlichungen, die im Text zitiert werden; eine an BEIER (1963) anschließende Literaturzusammenfassung findet sich bei MAHNERT (2004).

Familie Atemnidae KISHIDA, 1929

Gattung *Atemnus* CANESTRINI, 1884

Atemnus politus (E. SIMON, 1878)
   B

Familie Cheiridiidae H. J. HANSEN, 1893

Gattung *Apocheiridium* CHAMBERLIN, 1924

Apocheiridium ferum (E. SIMON, 1879)
   Ö (N, T)

Gattung *Cheiridium* MENGE, 1855

Cheiridium museorum (LEACH, 1817)
   Ö (N, O, T)

Familie Cheliferidae RISSO, 1826

Gattung *Chelifer* GEOFFROY, 1762

Chelifer cancroides (LINNÉ, 1758)
   Ö. Synanthrop; Freilandfunde unter Borke von Nadelbäumen beziehen sich
   höchstwahrscheinlich auf *Mesochelifer ressli* (s.u.).

Gattung *Dactylochelifer* BEIER, 1932

Dactylochelifer latreillei latreillei (LEACH, 1817)
   Ö (B, N)

Gattung *Hysterochelifer* CHAMBERLIN, 1932

Hysterochelifer meridianus (L. KOCH, 1873)
   St (eingeschleppt?)

V. Mahnert

Gattung *Mesochelifer* VACHON, 1940

Mesochelifer ressli MAHNERT, 1981
> O, N (loc. typ. = Lunz, Lunzberg), St, T. Bislang nur borkenbewohnend, von *Pinus* und *Picea*, bekannt. Freilandfunde von *Chelifer cancroides* sind wahscheinlich dieser Art zuzuordnen.

Gattung *Rhacochelifer* BEIER, 1932

Rhacochelifer peculiaris (L. KOCH, 1873)
> N (eingeschleppt)

Familie Chernetidae MENGE, 1855

Gattung *Allochernes* BEIER, 1932

Allochernes peregrinus (LOHMANDER, 1939)
> B* (Parndorfer Platte, Teichgraben), N, W* (Lobau)

Allochernes powelli (KEW, 1916)
> Ö (N, T)

Allochernes wideri wideri (C. L. KOCH, 1843)
> N, O, St, T

Gattung *Chernes* MENGE, 1855

Chernes cimicoides (FABRICIUS, 1793)
> Ö

Chernes hahnii (C. L. KOCH, 1839)
> B* (Parndorfer Platte, Teichgraben), N, T

Chernes montigenus (E. SIMON, 1879)
> T

Chernes nigrimanus (ELLINGSEN, 1897)
> K, N, St, T

Chernes similis (BEIER, 1932)
> St

Chernes vicinus (BEIER, 1932)
> K, N (loc. typ.= Kirling bei Wien: BEIER 1933, HELVERSEN 1966), O, St, T

Gattung *Dendrochernes* BEIER, 1932

Dendrochernes cyrneus (C. L. KOCH, 1873)
K, N, T

Gattung *Dinocheirus* CHAMBERLIN, 1929

Dinocheirus panzeri (C. L. KOCH, 1873)
K, N, T

Gattung *Lamprochernes* TÖMÖSVARY, 1882

Lamprochernes chyzeri (TÖMÖSVARY, 1882)
N, T* (Landeck)

Lamprochernes nodosus (SCHRANK, 1761)
Ö

Gattung *Lasiochernes* BEIER, 1932

Lasiochernes pilosus (ELLINGSEN, 1910)
B (loc. typ. = Goysz [heute Gois], Neusiedler See: BEIER 1929), K, N

Gattung *Pselaphochernes* BEIER, 1932

Pselaphochernes scorpioides (HERMANN, 1804)
Ö

Familie Chthoniidae DADAY, 1888

Gattung *Chthonius* C. L. KOCH, 1843

(Untergattung *Chthonius* s. str.)

Chthonius (C.) alpicola BEIER, 1951
K, O, S, St (loc. typ. = Preg bei Kraubath)

Chthonius (C.) carinthiacus BEIER, 1951
K (loc. typ. = Villacher Alpe: Warmbad Villach)

Chthonius (C.) ellingseni BEIER, 1939
K (loc. typ. = Annenheim am Ossiachersee)

V. Mahnert

Chthonius (C.) ischnocheles ischnocheles (HERMANN, 1804)
N, T

Chthonius (C.) jugorum BEIER, 1952
T

Chthonius (C.) orthodactylus (LEACH, 1817)
K, N, O, St

Chthonius (C.) pusillus BEIER, 1947
K, N (loc. typ. = Mitterdorf), St. Endemit.

Chthonius (C.) pygmaeus BEIER, 1934
K

Chthonius (C.) ressli BEIER, 1956
N (loc. typ. = Purgstall)

Chthonius (C.) submontanus BEIER, 1963
N (loc. typ. = Gaming), O, St. Endemit; alle außer-österreichischen Funde werden von GARDINI (2009) in Frage gestellt.

Chthonius (C.) tenuis L. KOCH, 1873
K, O, St

(Untergattung *Ephippiochthonius* BEIER, 1930)

Chthonius (E.) boldorii BEIER, 1934
K, O, S, St, T, V (Verbreitung nach MUSTER et al. 2004)

Chthonius (E.) fuscimanus E. SIMON, 1900
N, O, St, W (Verbreitung nach MUSTER et al. 2004)

Chthonius (E.) microtuberculatus HADŽI, 1937
B* (Parndorfer Platte, Zurndorfer Eichenwald)

Chthonius (E.) parmensis BEIER, 1963
B* (Parndorfer Platte, Zurndorfer Eichenwald), St

Chthonius (E.) tetrachelatus (PREYSSLER, 1790)
Ö (B, K, N, O, S, St, T)

Gattung ***Mundochthonius*** CHAMBERLIN, 1929

Mundochthonius alpinus BEIER, 1947
St (loc.typ. = Preg bei Kraubath). Endemit.

Mundochthonius styriacus Beier, 1971
    St (loc. typ. = Pöls bei Zwaring, W-Steiermark)

Familie Geogarypidae Chamberlin, 1930

Gattung *Geogarypus* Chamberlin, 1930

Geogarypus minor (L. Koch, 1873)
    N (eingeschleppt)

Familie Larcidae Harvey, 1992

Gattung *Larca* Chamberlin, 1930

Larca lata (H. J. Hansen, 1884)
    N

Familie Neobisiidae Chamberlin, 1930

Gattung *Microbisium* Chamberlin, 1930

Microbisium brevifemoratum (Ellingsen, 1903)
    K, St, T

Microbisium suecicum Lohmander, 1945
    N

Gattung *Neobisium* Chamberlin, 1930

(Untergattung *Neobisium* s. str.)

Neobisium (N.) caporiaccoi Heurtault-Rossi, 1966
    K

Neobisium (N.) carcinoides (Hermann, 1804)
    Ö

Neobisium (N.) carinthiacum Beier, 1939
    K (loc. typ. = Karawanken, Hochobir), St, T. Endemit.

Neobisium (N.) doderoi (E. Simon, 1896)
    K, St

V. Mahnert

Neobisium (N.) dolicodactylum (Canestrini, 1874)
K

Neobisium (N.) dolomiticum Beier, 1952
T, V. Alpines Höhenvorkommen, 1750-2550m.

Neobisium (N.) erythrodactylum (L. Koch, 1873)
B

Neobisium (N.) fuscimanum (C. L. Koch, 1843)
B, K, N, O, S, St, T

Neobisium (N.) galeatum Beier, 1953
K

Neobisium (N.) hermanni Beier, 1938
K, N (loc. typ. = Hermannshöhle b. Kirchberg am Wechsel), St, T. Nur aus Höhlen bekannt.

Neobisium (N.) jugorum (L. Koch, 1873)
T, V

Neobisium (N.) minimum (Beier, 1928)
N, St

Neobisium (N.) noricum Beier, 1939
S (loc. typ. = Pfandlscharte, Hohe Tauern). Nur vom Holotypus bekannt. Endemit.

Neobisium (N.) simile (L. Koch, 1873)
V

Neobisium (N.) simoni petzi Beier, 1939
N, O (loc. typ. = Feichtau-Seen, Sengsengebirge), St

Neobisium (N.) simoni simoni (L. Koch, 1873)
B, N, O, St

Neobisium (N.) sylvaticum (C. L. Koch, 1835)
B*, S, St, T

(Untergattung *Blothrus* Schiödte, 1847)

Neobisium (Blothrus) aueri Beier, 1962
O, St (loc. typ. = Almberg, Eis- und Tropfsteinhöhle bei Grundlsee, Grundlseer Berge). Troglobionte Art des Toten Gebirges. Endemit.

Gattung *Roncus* L. Koch, 1873

Roncus alpinus L. Koch, 1873
  K, T

Roncus carinthiacus Beier, 1934
  K (loc. typ. = Eggerloch b. Warmbad Villach). Nur aus wenigen Höhlen
  Kärntens bekannt. Endemit.

Roncus julianus Caporiacco, 1949
  T

Roncus lubricus (L. Koch, 1873)
  K (Vorkommen fragwürdig, siehe Problematica)

Roncus tenuis Hadži, 1933
  K* (Dobratsch)

Familie Syarinidae Chamberlin, 1930

Gattung *Syarinus* Chamberlin, 1925

Syarinus strandi (Ellingsen, 1901)
  N, T

Familie Withiidae Chamberlin, 1931

Gattung *Withius* Kew, 1911

Withius hispanus (L. Koch, 1873)
  N (eingeschleppt?)

Withius piger (E. Simon, 1878)
  Ö (T, N)

2. Problematica

Offensichtliche taxonomische Unsicherheiten betreffen drei Arten der Familie
Neobisiidae:

a. Stahlavsky et al. (2003) stellten bei *Neobisium carcinoides* das Auftreten von
    3 verschiedenenen Karyotypen fest (2n = 54, 60 oder 70, z.T. in sympatrischen
    Auftreten); dies deutet auf das Bestehen eines Arten-Komplexes hin, der jedoch

V. Mahnert

bislang morphologisch oder genetisch (DNS-Sequenzierung) nicht untersucht worden ist.

b. Die Neubeschreibung von *Roncus lubricus* L. Koch durch Gardini (1983) erbrachte den Beweis, dass auch hier ein Arten-Komplex vorliegt; eine Aufgliederung ist bislang nur teilweise erfolgt, die genaue Verbreitung von *R. lubricus* L. Koch ist unklar. Die Identität der aus Österreich (Kärnten) gemeldeten Exemplare ist unklar, eine Verwechslung mit *R. tenuis* Hadži wäre denkbar.

c. Bislang ist *Neobisium noricum* nur vom Holotypus bekannt; zusätzliche Exemplare vom locus typicus wären notwendig, um den Artstatus abklären zu können.

IV. Literatur

Beier, M. 1929: Die Pseudoskorpione des Wiener Naturhistorischen Museums. II. Panctenodactyli. — Ann. Nat.-Hist. Mus. Wien **43**: 341–367.

Beier, M. 1933: Revision der Chernetidae (Pseudoscorp.). — Zool. Jahrb., Syst. (Ökol.), Geogr. u. Biol. **64**: 509–548.

Beier, M. 1952: Pseudoscorpionidea, Afterskorpione. Catalogus Faunae Austriae **IX** a: 2-6. – Österr. Akad. Wiss., Springer-Verlag Wien.

Beier, M. 1956: Pseudoscorpionidea, Afterskorpione, 1. Nachtrag. Catalogus Faunae Austriae **IX** a: 8–9. — Österr. Akad. Wiss, Springer-Verlag Wien.

Beier, M. 1963: Ordnung Pseudoscorpionidea (Afterskorpione). Bestimmungsbücher zur Bodenfauna Europas **1**: i–vi+1–313. — Akad.-Verlag, Berlin.

Blick, T., Muster, Ch. & Duchac, V. 2004: Checkliste der Pseudoskorpione Mitteleuropas. Checklist of the pseudoscorpions of Central Europe. Version 1. Oktober 2004. — http://wwwAraGes.de/checklist/checklist04_pseudoscorpiones.html (eingesehen VIII. 2010)

Gardini, G. 1983: Redescription of *Roncus lubricus* L. Koch, 1873, type-species of the genus *Roncus* L. Koch, 1873. — Bull. Br. Arach. Soc. **6**: 78–82.

Gardini, G. 2009: *Chthonius (C.) delmastroi* n. sp. delle Alpi occidentali e del Piemonte e ridecrizione di *Chthonius (C.) tenuis* L. Koch, 1873 e di *C. (C.)subontanus* Beier, 1963 (Pseudoscorpiones Chthoniidae). — Riv. Piemont. Storia nat. **30**: 25–51.

Harvey, M.S. 1992: The phylogeny and classification of the Pseudoscorpionida (Chelicerata: Arachnida). — Invertebrate Taxonomy **6**: 1373–1435.

Harvey, M.S. 2009: Pseudoscorpions of the World. Version 1.2. Western Australian Museum, Perth. — http://www.museum.wa.gov.au/research/databases/pseudoscorpions (eingesehen VIII. 2010).

Helversen, O. von. 1966: Pseudoskorpione aus dem Rhein-Main-Gebiet. — Senckenbergiana Biologica **47**: 131–150.

JUDSON, M.L.I. 2010: A review of K. KISHIDA's pseudoscorpion taxa (Arachnida, Cheloneti). — Acta Arachnol. **59** (1): 9–13.

KOCH, L. 1873: Übersichtliche Darstellung der europäischen Chernetiden (Pseudoscorpione). — Verlag Bauer & Raspe, Nürnberg, 68 S.

KOCH, L. 1876: Verzeichnisse der in Tirol bis jetzt beobachteten Arachniden nebst Beschreibungen einiger neuen oder weniger bekannten Arten. — Zeitschrift des Museums Ferdinandeum Innsbruck (3) **20**: 219–354.

KOFLER, A. 1972: Die Pseudoskorpione Osttirols. — Mitt. zool. Ges. Braunau/Austria **1** (12): 286–289.

LENZ, H.O. 1856/1966: Zoologie der alten Griechen und Römer, deutsch in Auszügen aus deren Schriften, nebst Anmerkungen. — Dr. Martin Sändig oHG, Wiesbaden 656 S. (unveränderter Neudruck der Ausgabe von 1856).

MAHNERT, V. 2004: Die Pseudoskorpione Österreichs (Arachnida, Pseudoscorpiones). — Denisia **12**: 459–471.

MAHNERT, V. 2009: Pseudoscorpiones (Pseudoskorpione): 501-508. — In RABITSCH, W. & ESSL, F. : Endemiten - Kostbarkeiten in Österreichs Pflanzen- und Tierwelt. — Naturwissenschaftlicher Verein Kärnten und Umweltbundesamt GmbH, Klagenfurt und Wien, 924 S.

MUSTER, CH., SCHMARDA, TH. & BLICK, T. 2004: Vicariance in a cryptic species pair of European pseudoscorpions (Arachnida, Pseudoscorpiones, Chthoniidae). — Zool. Anz. **242**: 299–311.

RESSL, F. 1983: Die Pseudoskorpione Niederösterreichs mit besonderer Berücksichtigung des Bezirkes Scheibbs. — In RESSL, F. (Hrsg.): Naturkunde des Bezirkes Scheibbs. Die Tierwelt des Bezirkes Scheibbs **2**: 174–202.

RESSL, F. & BEIER, M. 1958: Zur Ökologie, Biologie und Phänologie der heimischen Pseudoskorpione. — Zool. Jahrb., Syst. (Ökol.), Geogr. u. Biol. **86** (1/2): 1–26.

SCHÜLLER, L. 1951: Beitrag zur Kenntnis der Pseudoscorpione im Lande Salzburg. — Mitt. Naturwiss. Arbeitsgem. Haus der Natur, Salzburg **2**: 1–9.

STAHLAVSKY, F., TUMOVA, P. & KRAL, J. 2003: Karyotype analysis in Central European pseudoscorpions of the genus *Neobisium* (Pseudoscorpiones: Neobisiidae). — Abstracts, 21st European Colloquium of Arachnology, St. Petersburg: 80.

Dr. Volker MAHNERT
c/o Muséum d'histoire naturelle
1, route de Malagnou, CH-1208 Genf, Schweiz
E-mail: volker.mahnert@wanadoo.fr

Tipulidae (Insecta: Diptera)

Peter VOGTENHUBER

Summary: This Checklist contents 141 species from the Family Tipulidae of Diptera, with known past or present occurence within present borders of Austria. In the world catalogue CCW (OOSTERBROEK 2010) at present, are also 141 species annouced from Austria, but three species had to be removed because their presence in Austria within the present borders is dubious – Austria in his old borders before 1918 was much more greater as nowadays. On the other side three species are added, who are discovered as new for the area. Against the species list given gy FRANZ (1990) much more material was collected in the meantime and it therefore gaves better knowledge of the distribution from the species within Austria. The distribution over the provinces will be given.

Zusammenfassung: Aus Österreich sind 141 Arten der Dipterenfamilie Tipulidae bekannt. Gegenüber dem Welt-Katalog (OOSTERBROEK 2010) ergab sich damit keine Änderung der Gesamtzahl, allerdings mussten drei Arten aus der Liste entfernt werden, die vermutlich in Österreich innerhalb der heutigen Grenzen nicht vorkommen. Andererseits kamen drei Arten hinzu, deren Vorkommen bisher nicht bekannt war. Gegenüber der ersten erschienen Liste (FRANZ 1990) wurde inzwischen viel gesammelt und es ergaben sich dadurch neue Erkenntnisse der Verbreitung über die einzelnen Bundesländer. Die derzeit bekannte Verbreitung bezogen auf die einzelnen Bundesländer wird angegeben.

Key Words: Diptera, Tipulidae, Austria, checklist, biodiversity

I Einleitung

Die Familie Tipulidae gehört zur Insektenordnung der Diptera (Zweiflügler), und zwar zur Unterordnung der Nematocera (Mückenartige). Derzeit sind weltweit 4388 Arten bzw. Unterarten von Tipulidae bekannt (OOSTERBROEK 2010), in Europa gibt es etwa 470 und in Österreich 141 Arten. Sie bilden zusammen mit den Limoniidae, Pediciidae und Cylindrotomidae die Superfamilie der Tipuloidea. Die einzige deutschsprachige Bezeichnung ist "Schnaken", diese grenzt sie aber nicht sicher von den anderen Mücken ab, denn je nach Gegend werden damit oft auch Stechmücken bezeichnet. Die auch in Österreich vorkommende *Tipula maxima* hat eine Flügelspannweite von 60 mmn und ist eine der größten europäischen Tipuliden.

Die Larven der meisten Arten ernähren sich von totem Pflanzenmaterial, wie Laub und Nadeln oder morschem Holz, nur wenige gehen auch an lebendes Pflanzenmaterial wie die Arten des Subgenus *Tipula*, die auch die Wurzeln von Gräsern und Wiesenkräutern fressen und so auf Rasenflächen schädlich werden können. Der überwiegende Teil lebt terrestrisch in den obersten Bodenschichten, einige semiaquatisch im Boden von Feuchtwiesen oder Flachmooren und einige aquatisch am Grund von Gewässern – besonders jene des Subgenus *Yamatotipula*.

Tipulidae haben nach heutigem Erkenntnisstand ihren Schwerpunkt nicht in Gebieten mit tropischem Klima, sondern erreichen die größten Artenzahlen in Zonen mit gemäßigtem Klima.

Danksagung: Ich danke dem Kustoden der Dipterensammlung des Naturhistorisches Museum Wien (NMW), Herrn Peter SEHNAL, für die Möglichkeit die Sammlung durchzusehen; ebenso dem Leiter des Museums im Stift Admont, Dr. Gerald UNTERBERGER, und dem Kustoden der STROBL-Sammlung Prof. CHVALA dafür, dass ich die Tipuliden untersuchen konnte.

Weiters danke ich allen, die mir ihre Tipulidenfänge gaben, und auch jenen, die mir ihre Tipulidenaufsammlungen zur Determination und Datenerfassung überließen. Es sind dies: Josef GUSENLEITNER, Ejulf und Ulrich AISTLEITNER, Fritz GUSENLEITNER, Franz RESSL, Ernst HÜTTINGER, Hubert und Renate RAUSCH, Hans MALICKY, Franz LICHTENBERGER, Theodor KUST, Gerhard TARMANN, Alois KOFLER, Franz SPETA, Martin SCHWARZ, Andreas LINK, Fritz LAUBE und Norbert PÖLL. Weiters bekam ich Material von †Dr. Karl BURMANN, †Prof. Dr. Konrad THALER und †Dr. Anton ADLMANNSEDER.

II Allgemeiner Teil

1. Geschichte der Erforschung der österreichischen Tipuliden

LINNÉ sowie FABRICIUS beschrieben schon Tipuliden von denen auch viele in Österreich zu finden sind. Von den in auch in Österreich vor 1800 tätigen Naturwissenschaftlern ist vor allem Nikolaus PODA VON NEUHAUS (1723–1798) zu nennen, der in seiner "Insecta Musei Gracensis" (1761) *Tipula maxima* beschrieb. Von überragender Bedeutung für die gesamte Dipterenforschung war der deutsche Entomologe Johann Wilhelm MEIGEN (1764–1845).

Im 19. Jhd. wirkten in Österreich: Johann Georg EGGER (1804–1866), Ignatz Rudolf SCHINNER (1813–1873), JOSEF MIK (1839–1900), Emanuel POKORNY (1837–1900), Pater Gabriel STROBL (1846–1925).

Und im 20. Jhd.: Leander CZERNY (1859–1944), Bernhard MANNHEIMS (1909–1971), Theowald VAN LEEUWEN (1919–2003), Ernst PECHLANER (1901–1964) und MANNHEIMS (1963) lieferten die erste Gebietsmonographie eines österreichishen Bundeslandes (Tirol) Charles Paul ALEXANDER (1889–1981); Günter THEISCHIGER erstellte die erste Gebietsmonographie Oberösterreichs, er lebt seit 1979 in Australien.

2. Datengrundlage

Ein Großteil des Materials befindet sich dzt. in der Sammlung des Biologiezentrums der OÖ. Landesmuseen in Linz und wurde großteils nach 1970 von verschiedenen Personen gesammelt. Weiters wurden durchgesehen: Die Sammlung im NMW und die Sammlung STROBL im Stift Admont.

Einige Funde konnte ich nicht besichtigen, so waren viele von H. Franz (1989 u. 1990) angegebene und von ihm gesammelte Tipuliden nicht auffindbar, allerdings wurde ein großer Teil seiner Aufsammlungen nach seinen Angaben von Mannheims determiniert. Die Sammlung PECHLANER in Innsbruck wurde nicht durchgesehen, da alles von Mannheims determiniert wurde, es mussten bloß die nomenklatorischen Änderungen berücksichtigt werden.

3. Aktueller Durchforschungsstatus

Der Kenntnisstand ist in den einzelnen Bundesländern und auch innerhalb der Bundesländer stark unterschiedlich. Am besten ist Oberösterreichs Fauna erforscht. Von Niederösterreich ist der Südwesten intensiv besammelt worden, vom übrigen Bundesland. ist die Tipulidenfauna weniger bekannt. Im Burgenland ist nur die Fauna der Umgebung des Neusiedlersees einigermaßen untersucht.

Von der Steiermark ist der Nordteil etwas besser besammelt, jedoch die Tipuliden der Süd- und Oststeiermark sind nahezu unbekannt.

In Kärnten ist die Datendichte mäßig. Von Salzburg gibt es nur wenige unregelmäßig verteilte Fundorte. Nordtirol weist eine gute Durchforschung auf, von Osttirol gibt es jedoch weniger Daten. Vorarlberg ist nur in den Lagen unter 1000 m einigermaßen gut besammelt.

Gebietseinteilung

B	=	Burgenland
K	=	Kärnten
N	=	Niederösterreich,Wien
O	=	Oberösterreich
Ot	=	Osttirol (Nordtirol ist extra ausgewiesen)
S	=	Salzburg
St	=	Steiermark
T	=	Nordtirol
V	=	Vorarlberg
Ö	=	Österreich, aus allen Bundesländern gemeldet

Abkürzungen:

*	=	Neufund in Österreich, bisher nicht schriftlich veröffentlicht
loc.typ.	=	Locus typicus (Originalfundort)
syn	=	synonym

III Spezieller Teil

1. Liste der in Österreich nachgewiesenen Arten und Unterarten

Familie Tipulidae

Genus *Dictenidia* BRULLÉ, 1833

Dictenidia bimaculata (LINNAEUS, 1761)
   B, N, O, St, T

Genus *Dolichopeza* CURTIS, 1825

Subgenus *Dolichopeza* CURTIS, 1825

Dolichopeza albipes (STRÖM, 1768)
N, O, St

Dolichopeza nitida MIK, 1874
N, S

Genus *Ctenophora* MEIGEN 1803

Subgenus *Cnemoncosis* ENDERLEIN 1921

Ctenophora festiva MEIGEN, 1804
N, O, St

Ctenophora ornata MEIGEN, 1818
K, N, O, Ot

Subgenus *Ctenophora* MEIGEN 1803

Ctenophora elegans MEIGEN, 1818
N, O, St

Ctenophora flaveolata (FABRICIUS, 1794)
N, O, S, St, T

Ctenophora guttata MEIGEN, 1818
N, S, St

Ctenophora pectinicornis (LINNAEUS, 1758)
N, O, St, T

Genus *Nephrotoma* MEIGEN, 1803

Nephrotoma aculeata (LOEW, 1871)
loc.typ. "Austria"; K, N, O, Ot, S, St, T

Nephrotoma analis (SCHUMMEL, 1833)
B, K, N, O, St, T

Nephrotoma appendiculata appendiculata (PIERRE, 1919)
B, K, N, O, St, T (als *Pales maculata*), V

Nephrotoma austriaca (MANNHEIMS & THEOWALD, 1959)
Ot, S, St (loc.typ.= Enns Au bei Schladming,), T

Nephrotoma cornicina cornicina (LINNAEUS, 1758)
  K, N, O, Ot, S, St, T, V

Nephrotoma crocata crocata (LINNAEUS, 1758)
  B, K, N, O, Ot, St, T

Nephrotoma croceiventris lindneri (MANNHEIMS, 1951)
  B, N, O, Ot, St, T

Nephrotoma dorsalis (FABRICIUS, 1782)
  N, O, S, T, V

Nephrotoma flavescens (LINNAEUS, 1758)
  K, N, O, S, St, T, V.

Nephrotoma flavipalpis (MEIGEN, 1830)
  N

Nephrotoma lunulicornis (SCHUMMEL, 1833)
  K, N, O, St, V

Nephrotoma pratensis pratensis (LINNAEUS, 1758)
  N, O, Ot, St, T

Nephrotoma quadrifaria quadrifaria (MEIGEN, 1804)
  N, O, S, St, V

Nephrotoma quadristriata (SCHUMMEL, 1833)
  N, O, S, St, T

Nephrotoma scalaris scalaris (MEIGEN, 1818)
  B, N, O, St

Nephrotoma scurra (MEIGEN, 1818)
  K, N, O, S, St, T

Nephrotoma submaculosa EDWARDS, 1928
  B, N, O

Nephrotoma tenuipes (RIEDEL, 1910)
  K, N, O, S, St, T, V

Genus *Nigrotipula* HUTSON & VANE-WRIGHT, 1969

Nigrotipula nigra nigra (LINNAEUS, 1758)
  B, K, N, O, Ot, T

P. Vogtenhuber

Genus *Prionocera* LOEW, 1844

Prionocera pubescens LOEW, 1844
S

Genus *Taniptera* LATREILLE, 1804

Taniptera atrata atrata (LINNAEUS, 1758)
Ö

Taniptera nigricornis nigricornis (MEIGEN, 1818)
K, N, O, S, St

Genus *Tipula* LINNEAUS

Subgenus *Acutipula* ALEXANDER

Tipula (A.) bosnica STROBL 1998
syn. *Tipula decipiens* CZIZEK
B, N, O, T.

Tipula (A.) fulvipennis DE GEER 1776
K, N, O, S, T, V

Tipula (A.) luna WESTHOFF 1879
B, K, N, O, St, T

Tipula (A.) maxima PODA 1761
N, O, S, St, T

Tipula (A.) tenuicornis SCHUMMEL 1833
B, N, O

Tipula (A.) vittata MEIGEN 1804
N, O, St

Subgenus *Beringotipula* SAVCHENKO, 1961

Tipula (B.) unca WIEDEMANN, 1871
K, N, O, S, St

Subgenus *Dendrotipula* SAVCHENKO, 1964

Tipula (D.) flavolineata MEIGEN, 1804
K, N, O, St

Subgenus *Emodotipula* ALEXANDER, 1966

Tipula (E.) obscuriventris STROBL, 1900
   N, O, V

Tipula (E.) saginata BERGROTH, 1891
   O, S, St, T

Subgenus *Lindnerina* MANNHEIMS, 1965

Tipula (L.) bistilata bistilata LUNDSTRÖM, 1907
   N, O, St, T

Subgenus *Lunatipula* EDWARDS, 1931

Tipula (L.) adusta adusta SAVCHENKO, 1954
   T

Tipula (L.) affinis SCHUMMEL, 1833
   O, T

Tipula (L.) alpina LOEW, 1873
   K (loc.typ.= "Kärntner Alpen"), N, O, T

Tipula (L.) borysthenica SAVCHENKO, 1954
   B

Tipula (L.) bullata LOEW, 1873
   loc.typ.= "Kärnten, Steiermark"; K, S, St, T

Tipula (L.) circumdata SIEBKE, 1863
   K, Ot, St, T

Tipula (L.) fascingulata MANNHEIMS, 1966
   B, K, N, O, St, T

Tipula (L.) fascipennis MEIGEN, 1818
   K, N, O, S, St, T

Tipula (L.) helvola LOEW, 1873
   B, K, N, O, St, V

Tipula (L.) laetabilis ZETTERSTEDT, 1838
   N, O, S, St, Ot, T

Tipula (L.) limitata SCHUMMEL, 1833
   K, O, Ot, S, St, T

P. Vogtenhuber

Tipula (L.) livida livida VAN DER WULP, 1859
  K, O, Ot, St, T

Tipula (L.) lunata LINNAEUS, 1758
  B, N, O, S, T

Tipula (L.) magnicauda STROBL, 1859
  N, O, St (loc.typ.= bei Rottenmann), T

Tipula (L.) mellea SCHUMMEL, 1833
  N

Tipula (L.) pannonia pannonia LOEW, 1873
  B

Tipula (L.) peliostigma peliostigma SCHUMMEL, 1833
  B, N, O

Tipula (L.) pokornyi MANNHEIMS, 1968
  (loc.typ.= Leitha Gebirge)

Tipula (L.) recticornis SCHUMMEL, 1833
  O, T

Tipula (L.) selene MEIGEN, 1830
  B, N, O, St

Tipula (L.) soosi soosi MANNHEIMS, 1954
  B, N

Tipula (L.) truncata truncata LOEW, 1873
  B, N

Tipula (L.) vernalis MEIGEN, 1804
  B, K, N, O, St, T, V

Subgenus *Mediotipula* PIERRE, 1924

Tipula (M.) mikiana BERGROTH, 1888
  K (loc.typ.= Mallnitz), Ot, S, St, T

Tipula (M.) sarajevensis STROBL, 1898
  S, St, T

Tipula (M.) siebkei ZETTERSTEDT, 1852
  N, St

Tipula (M.) stigmatella SCHUMMEL, 1833
N, O, T

Subgenus *Odonatisca* SAVTSHENKO 1956

Tipula (O.) nodicornis MEIGEN 1818
N, O

Subgenus *Platytipula* MATSUMURA 1916

Tipula (P.) luteipennis MEIGEN 1830
B, K, N, O, S, St, T

Tipula (P.) melanocerus SCHUMMEL 1833
N, O, St

Subgenus *Pterelachisus* RONDANI

Tipula (P.) austriaca POKORNY 1887
N (loc.typ.= Schneeberg), O, Ot, St, T

Tipula (P.) bilobata POKORNY 1887
K, N, O, S, St, T

*****Tipula (P.) cinereocincta*** LUNDSTRÖM, 1907
N oder St. (ein Männchen in der Sammlung STROBL/Admont mit zwei
Fundortangaben!), T

Tipula (P.) crassiventris RIEDEL, 1913
St (alle aus den Nordalpen bisher untersuchten Exemplare haben sich als *T.
pseudocrassiventris* erwiesen!)

Tipula (P.) glacialis POKORNY, 1887
O, S, T, V

Tipula (P.) irregularis POKORNY, 1887
K, Ot, S, St, T

Tipula (P.) irrorata MACQUART, 1826
K, N, O, St, T

Tipula (P.) luridorostris SCHUMMEL, 1833
O, S, St

Tipula (P.) mayerduerii EGGER,1863
K, N, O, Ot, St, T

P. Vogtenhuber

Subgenus *Schummelia* EDWARDS

Tipula (S.) variicornis variicornis SCHUMMEL 1833
N, O, S, St, T

Tipula (S.) zernyi THEOWALD, 1952
K (loc.typ.=Birnbaum, Gailtaler Alpen), N, O, S, St, T

Tipula (S.) zonaria GOETGHEBUER, 1921
N, O, T

Subgenus *Tipula* LINNEAUS

Tipula (T.) hungarica LACKSCHEWITZ, 1930
Zweifelhafte Art, eventuell Hybrid *T. oleracea x orientalis?*
B (loc.typ.= Weiden, Seewinkel)

Tipula (T) mediterranea LACKSCHEWITZ, 1930
N

Tipula (T) oleracea LINNEAUS, 1758
B, N, O, S, St, T, V

Tipula (T) orientalis LACKSCHEWITZ, 1930
B, N

Tipula (T) paludosa MEIGEN, 1803
Ö

Tipula (T) subcunctans ALEXANDER, 1921
B, N, Ot, T

Subgenus *Vestiplex* BEZZI, 1924

Tipula (V.) crolina DUFOUR 1992
Früher *carolae* DUFOUR (Homonym)
K, Ot, S, T

Tipula (V.) excisa excisa SCHUMMEL 1833
K, N, O, Ot, S, St, T

Tipula (V.) franzi MANNHEIMS 1950
S, St (loc.typ.= Schladminger Tauern, E Deichselspitze)
Österreich – Endemit

Tipula (V.) hemiptera strobliana MANNHEIMS 1966
syn. *cinerea* STROBL (1895)
O, S, St (loc.typ.= Gesäuse, Kalbling), T, V

Tipula (V.) hortorum L. 1758
B, K, N, O, S, St, T

Tipula (V.) montana montana Curtis 1834
K, N, O, Ot, S, St, T

Tipula (V.) nubeculosa Meigen 1804
B, K, N, O, Ot, St, T, V

Tipula (V.) pallidicosta pallidicosta Pierre 1824
N, O, Ot, S, St, T

Tipula (V.) scripta scripta Meigen 1830
K, N, O, Ot, S, St, T, V

Tipula (V.) sexspinosa Strobl 1898
K, St (loc.typ.= Koralm)
Österreich – Endemit

Subgenus *Yamatotipula* Matsumura

Tipula (Y.) afriberia italia Theowald, Dufour & Oosterbroek, 1982
N, O

Tipula (Y.) caesia Schummel, 1833
K, O, St, T

Tipula (Y.) coerulescens Lackschewitz, 1923
O

Tipula (Y.) couckei Tonnoir, 1921
O, T, V

Tipula (Y.) lateralis Meigen, 1804
B, K, N, O, S, St, T, V

Tipula (Y.) marginella Theowald, 1980
K, O, St

Tipula (Y.) montium Egger, 1863
K, N, O (loc.typ.= Gmunden), T

Tipula (Y.) pierrei Tonnoir, 1921
syn. *solstitialis* Westhoff 1879
N, O, T

P. Vogtenhuber

Tipula (Y.) pruinosa pruinosa WIEDEMANN, 1817
    K, N, O, S, St, T, V

Tipula (Y.) riedeli MANNHEIMS, 1952
    St

Tipula (Y.) submontium THEOWALD & OOSTERBROEK, 1981
    N

2. Problematica

2.1. Arten deren Vorkommen in Österreich zweifelhaft ist: (Diese Arten sind nicht in der vorangegangenen Liste angeführt)

Prionocera subserricornis (ZETTERSTEDT, 1851)
    Vorkommen ungewiss, nur ein Exemplar im NMW mit „leg. SCHINNER", ohne sonstige Daten.

Prionocera turcica (FABRICIUS, 1787)
    Vorkommen ungewiss, nur ein Exemplar im NMW „leg. EGGER", ohne sonstige Daten.

Tipula (Yamatotipula) fenestrella THEOWALD, 1980
    Das vermutete Vorkommen in Österreich begründet sich auf ein Exemplar im NMW mit dem Etikett "Austria, alte Sammlung, SCHINNER 1869". Damals bezog sich aber "Austria" auf die Grenzen der Habsburgischen Monarchie mit der Nord – Süd Ausdehnung von österr. Schlesien (heute Polen) bis Dalmatien. Diese Art wurde aus Tschechien, Polen und weiter nordwärts gemeldet.

Tipula (Pterelachisus) matsomuriana pseudohortensis LACKSCHEWITZ, 1932
    Früher *Tipula (Pterelachisus) pseudohortensis* LACKSCH. Fundort OÖ. Freistadt (FRANZ 1989); THEOWALD (1980) vermutet Fehlbestimmung, weil die Art nur von Lettland, Estland, Finnland und Russland bekannt ist. Wird im Katalog von FRANZ (1990) noch angeführt, jedoch nicht mehr bei OOSTERBROEK (2010).

2.2. Arten die seit längerer Zeit nicht wieder gefunden wurden: (Diese Arten sind in der vorangegangenen Liste angeführt)

Ctenophora elegans MEIGEN, 1818: Wurde seit 60 Jahren nicht mehr gefunden.
Tipula winthemi LACKSCHEWITZ, 1932: Darüber schreibt THEOWALD (1980): "Die meisten Funde stammen von vor mehr als 100 Jahren." An neueren Funden führt er an: 1928 (1Mn. an der Mündung des Ob), 1929 (1Wb. Kamschatka), 1947,

1953 u. 1967 (insges. 2Mn., 2Wb. Ardennen u. Eifel). Aus Österreich ist mir kein rezentes Exemplar bekannt. Die Erstbeschreibung erfolgte nach einem Museumsexemplar; LACKSCHEWITZ lagen Exemplare aus Gmunden, Freistadt und Biglerhütte (N) vor.

IV Literatur

BRODO F. 1987: A revision of the genus *Prionocera* (Diptera: Tipulidae). — Evol. Monogr., **8**: 1–93.

DUFOUR, C. 1983: *Tipula (Savtshenkia) tulipa* sp. n. from xerothermic valleys of the Swiss Alps (Diptera: Tipulidae). — Mitt. schweiz. ent. Ges., **56**: 275–281.

DUFOUR, C. 1984: *Tipula (Vestiplex) carolae* sp. n., a high alpine species of the excisa group (Diptera: Tipulidae). — Mitt. schweiz. ent. Ges., **57**: 79–84.

DUFOUR, C. 1986: Les Tipulidae de Suisse (Diptera, Nematocera). — Doc. Faun. Helvetiae, **2**: 1–187, Tafeln 1–149.

DUFOUR, C. 1991: The identity of *Tipula (Emodotipula) saginata* BERGROTH and *T. (E.) obscuriventris* STROBL, and the description of *Tipula (E.) leo* sp. n. from the Sierra Nevada in Spain (Diptera, Tipulidae). — Mitt. schweiz. ent. Ges., **64**: 81–91.

FRANZ, H. 1943: Die Landtierwelt der mittleren Hohen Tauern. — Denkschrift Akad. Wiss. Wien. **107**: 1–552.

FRANZ, H. 1989: Diptera Orthorapha. In: Die Nordost-Alpen im Spiegel ihrer Landtierwelt. — Innsbruck, **6**(1): 1–413.

FRANZ, H. 1990: Fam.: Tipulidae, Limoniidae, Cylindrotomidae, Ptychopteridae. — Catalogus Faunae Austriae **19a**: 1–57.

MANNHEIMS, B. 1951–1968: 15. Tipulidae. — In LINDNER, E. (Ed.): Die Fliegen der palaearktischen Region, **3**(5)1: 1–320

MANNHEIMS, B. & PECHLANER, E. 1963: Die Tipuliden Nordtirols (Dipt.). – Stuttg. Beitr. Naturk., **102**: 1–29.

OOSTERBROEK, P. 1978–1980: The western palaearctic species of *Nephrotoma* MEIGEN, 1803: (Diptera,Tipulidae). — Beaufortia, **27**: 1–137, **28**: 57–111, **28**: 157–203, **29**: 1–137, **29**: 311–393.

OOSTERBROEK, P.; THEOWALD, BR. 1992: Family Tipulidae. — Catalogue of Palaearctic Diptera **1**: 56–178.

OOSTERBROEK, P. 1994: Notes on western Palaearctic species of the *Tipula (Yamatotipula) lateralis* group, with the description of a new species from Turkey (Diptera: Tipulidae). — European Journal of Entomologie **91**: 429–435.

OOSTERBROEK, P. 2010: Catalogue of the Craneflies of the World (CCW). — Erhältlich unter http.//ip30.eti.uva.nl/ccw/ (last update: 9. September 2010).

P. Vogtenhuber

OOSTERBROEK, P.; BYGEBJERG, R. & MUNK, T. 2006: The West Palaearctic species of Ctenophorinae (Diptera: Tipulidae), key, distribution and references. — Ent. Ber., Amst. **66**(5): 138–149.

SAVTSHENKO, E. N. & THEISCHINGER, G. 1978: Die Arten der *Tipula (Lunatipula) recticornis* – Gruppe (Diptera, Tipulidae). — Bull. Zool. Mus. Univ. Amsterdam, **6**: 117–128.

STROBL, G. 1893: Die Dipteren von Steiermark, 3. Theil. — Mitt. Naturwiss. Ver. Steiermark, **31**: 121–246.

STROBL, G. 1897: Die Dipteren von Steiermark, 4. Theil: Nachträge. — Mitt. Naturwiss. Ver. Steiermark, **31**: 191–298.

STROBL, G. 1900: Tief's dipterologischer Nachlass aus Kärnten und Oesterr.-Schlesien. — Jb. naturh. Landesmus. Kärnten, 26: 171–246.

STROBL, G. 1909: Die Dipteren von Steiermark, 2. Nachtrag. — Mitt. Naturwiss. Ver. Steiermark, **46**: 45–293.

THEISCHINGER, G. 1978: Schnaken (Tipulidae) aus Oberösterreich (I), (Diptera, Nematocera). — Jb. Oö. Mus.-Ver. **123**: 237–268.

THEISCHINGER, G. 1980: Schnaken (Tipulidae) aus Oberösterreich (II), (Diptera, Nematocera). — Jb. Oö. Mus.-Ver. **125**: 251–254.

THEOWALD, BR. 1973 u. 1980: 15. Tipulidae. — In LINDNER, E. (Ed.): Die Fliegen der palaearktischen Region, 3(5)1: 321–538.

THEOWALD, BR. & MANNHEIMS, B. 1962: Die Arten der *Tipula (Vestiplex) excisa*-Gruppe in der Paläarktis (Diptera, Tipulidae). — Bonn. Zool. Beitr. **13**: 360–402.

VERMOOLEN, D. 1983: The *Tipula (Acutipula) maxima* group (Insecta, Diptera, Tipulidae), I. Taxonomy and Distribution. — Bijdr. Dierk., **53**(1): 49–81.

VOGTENHUBER P. 1994: Neue und bemerkenswerte Tipuliden aus Oberösterreich (Insecta: Diptera: Tipulidae) — Beitr. Naturk. Oberösterreichs **2**: 175–186.

VOGTENHUBER P. 1996: Zwei für Oberösterreich neue Tipuliden-Arten (Insecta: Diptera: Tipulidae) — Beitr. Naturk. Oberösterreichs **4**: 49–51.

VOGTENHUBER P. 2002: *Tipula (Vestiplex) sexspinosa* STROBL, 1897 nicht nur ein Endemit der Koralpe (Diptera: Tipulidae). — Carinthia II(192) **112**: 541–544.

VOGTENHUBER P. 2004: Bemerkenswerte Tipulidenfunde aus Oberösterreich (Insecta: Diptera: Tipulidae) — Beitr. Naturk. Oberösterreichs **13**: 407–411.

VOGTENHUBER P. 2009: Dipteren (Fliegen und Mücken). — In RABTISCH, W. & ESSL, F. (Eds.): Endemiten – Kostbarkeiten in Österreichs Pflanzen- und Tierwelt. — Naturwissenschaftlicher Verein für Kärnten und Umweltbundesamt GmbH, Klagenfurt und Wien: 790 – 795.

Dipl.-Ing. Peter VOGTENHUBER
Biologiezentrum der Oberösterreichischen Landesmuseen
J.-W.-Klein-Str. 73, A-4040 Linz/Dornach, Austria
E-Mail: p.vogtenhuber@landesmuseum.at